AN INTRODUCTION TO PLASMA ASTROPHYSICS AND MAGNETOHYDRODYNAMICS

ASTROPHYSICS AND SPACE SCIENCE LIBRARY

VOLUME 294

AN INTRODUCTION TO PLASMA ASTROPHYSICS AND MAGNETOHYDRODYNAMICS

by

MARCEL GOOSSENS
Centre for Plasma Astrophysics,
K.U. Leuven, Belgium

SPRINGER-SCIENCE+BUSINESS MEDIA, B.V.

A C.I.P. Catalogue record for this book is available from the Library of Congress.

ISBN 978-1-4020-1433-8 ISBN 978-94-007-1076-4 (eBook)
DOI 10.1007/978-94-007-1076-4

Printed on acid-free paper

Contents

Preface

This book aims to give a basic introduction to plasma physics with an emphasis on Magnetohydrodynamics (MHD for short). It has grown out of the lecture notes that I have been teaching at the K.U.Leuven for the last fifteen years to third and fourth year undergraduate students in applied mathematics and physics. For students at the K.U.Leuven this course is their first encounter with plasma physics. Hence, its level is elementary. Since both students in mathematics and physics can take this course, both mathematical integrity and physical intuition are pursued.

The content of this book might not be what is routinely expected from an introductory book on plasma physics. Often introductory courses present various approximate plasma models with minimal discussion of the underlying theoretical foundations and the simplifying assumptions. In contrast, I have decided to give special attention to fundamental concepts and first principles and to limit the discussion of plasma models to a large extent to Magnetohydrodynamics (MHD). The equations of MHD are derived from first principles starting from the Boltzmann equations for the distribution functions in phase space of the various species. It is clear that this approach is not the best way to economically present the equations of MHD. However, it has the considerable advantage that it allows to show where and how microphysics has been removed and to determine the limitations and the domain of validity of MHD. I have found out that an elementary level and a concern with fundamental concepts and first principles are not mutually exclusive, but go together very well. The reader will not end up with the wrong impression that MHD covers all of plasma physics. He/she will be fully aware that plasmas can show behaviour that cannot be captured with MHD. In that sense the course is also an invitation to the interested reader to go beyond classic MHD and discover many fascinating plasma behaviour that is not dealt with here.

I decided to set this course in the framework of solar and space plasma physics and astrophysics. Hence, the name of the course: "Introduction to Plasma Astrophysics and Magnetohydrodynamics". The fully and partially ionized plasmas that are the central focus of solar and space physics are related on a fundamental level to laboratory plasma physics, which directly investigates basic plasma physical processes, and to astrophysics, a discipline that relies heavily on understanding the physics of the plasma state. Although solar physics is the framework in which I want to set my course, I have refrained myself from concentrating on a description of a large number of plasma physics phenomena in solar physics and astrophysics. The first chapter gives an incomplete and short overview of plasma physics phenomena in solar physics and astrophysics. The last chapter is the only chapter that deals exclusively with solar physics. It discusses the solar wind in the context of hydrodynamics and MHD.

The amount of mathematics and physics required for using this book is limited. A knowledge of vector calculus, real calculus and electromagnetic theory are the modest prerequisites from mathematics and physics. The exercises take a special place in the course and in this book. We all learn best the things that we have discovered for ourselves. Hence, in stead of being very detailed in the derivation of the equations and results, I have taken the relaxed attitude to be economical on intermediate results and steps when these are straightforward. A first class of exercises invites the reader to fill in gaps in the often long derivation of equations. There is no need for the reader to memorize the different steps required for obtaining a given equation or result, but he/she should have gone through this straightforward mathematics at least once. Also, it is difficult, if not impossible, for anyone to learn a subject purely by reading about it. Applying the information to specific problems and thereby being encouraged to think about what has been read, is essential in the learning process. A second class of exercises tries to invite the reader to just do that. The exercises form a major part of this book. In the Belgian educational system students have to take exams. The exam for the material covered in this book consists of solving the exercises and explaining the solutions by using the notes the students have prepared themselves.

A short and elementary book on "Plasma Astrophysics and Magnetohydrodynamics" cannot focus on recent research results which require a deep understanding of the subject. Even at this elementary level I feel that the insights and the interpretations that I try to convey in this book, are influenced and shaped to a large extent by the scientific collaborations and discussions I have had over the years with numerous colleagues and friends including in particular L. Mestel, Z. Sedlacek, A.D.M. Walker, J.P. Goedbloed, F. Verheest, E.R. Priest, J.V. Hollweg, W. Kerner, B. R. Roberts, M. Ruderman, T. Sakurai, K. Tsinganos, A. W. Hood, Y. Voitenko, D. Van Eester, S. Poedts, R. Erdelyi, R. van der Linden and R. Keppens. In addition to being instructive, it was fun. Thank you. My gratitude also goes to A. De Groof for her help in preparing this book and to P. Charbonneau for providing me with ps-files of figures of his unpublished class notes on "Large Scale Dynamics of the Solar Wind". It is a pleasure to thank Kluwer Academic Publishers for giving me the opportunity to publish my class notes in the Astrophysics and Space Science Library Series. I have benefitted from several good books on plasma physics, magnetohydrodynamics and solar physics. Those that I like the best are listed at the end of the introductory chapter under references. These are, with one or two exceptions, the only references given in this book. The material covered in this book is at the basic elementary level and owned as it were by the community. It should be clear that nothing of the work described in this book is my own. The book is based on the work by the pioneering giants J.C. Maxwell, L. Boltzmann, I. Langmuir, J. Larmor, H. Alfvén and E. Parker.

The students who have taken this course over the years, probably do not realize it, but I have benefitted a lot from them. Their criticism and questions have helped me shaping the notes in their present form. The fact that several of these former students are pursuing scientific and academic careers in which mathematical modelling of plasmas and MHD still play a prominent role, is reassuring to me. It has not all been in vain. Most of the students who have taken the course, do not use its content in their daily professional life. For them I dare hope that the course has contributed to their scientific training by learning them how mathematical modelling and physical intuition and interpretation can go hand in hand. The importance of mathematical modelling in this context must be stressed. Even when the mathematical description has been simplified by replacing a description using the Boltzmann equations for the distribution functions in phase space of the various species, with a description based on MHD, mathematical modelling is often essential. The full nonlinear equations of

MHD are so complicated, that they often need to be approximated drastically by focussing on the dominant physical mechanisms in a any particular situation. When solutions for simple situations are known, more and more effects may be added to make the model more realistic. It is my hope that this book may help students at the K.U.Leuven and elsewhere to appreciate the intriguing and complicated behaviour of plasmas and to appreciate the power of mathematical modelling as a tool for exploring and understanding this complicated behaviour.

March 2003

Marcel Goossens

Figure 1: *Cover Illustration.* (Courtesy of SOHO (ESA & NASA.))

Cover illustration

This cover illustration is a composite of

- EIT EUV image taken in the Fe XV line at 284A showing the corona above the disk at a temperature of about 2-2.5 million K (innermost image)
- UVCS image showing the Sun's outer atmosphere as it appears in ultraviolet light emitted by electrically charged oxygen (O VI) flowing away from the Sun to form the solar wind (middle region), and
- Image of the extended white light corona as recorded by the outer LASCO coragraph (C3) on 23 December 1996.

The field of view of this instrument encompasses 32 diameters of the Sun. To put this in perspective, the diameter of this image is 45 million kilometers at the distance of the Sun, or half of the diameter of the orbit of Mercury. During that time of the year, the Sun is located in the constellation Sagittarius. The center of the Milky Way is visible, as well as the dark interstellar dust rift, which stretches from the south to the north. Three prominent streamers can be seen (two at the West and one at the East limb). This image also shows

Comet SOHO-6 (elongated streak at about 7:30 hours), one of several tens of sun-grazing comets discovered so far by LASCO. It eventually plunged into the Sun.

This composite image can be found at http://sohowww.nascom.nasa.gov/gallery/LASCO/ (Courtesy of SOHO (ESA & NASA.))

Chapter 1

Introduction

> "That is why I love elementary school so much.
> The kids really believe everything you tell them."
> *Principal Seymour Skinner to Mrs. Edna Crabapple.*
> The Simpsons.

This course is concerned with plasma physics with an emphasis on Magnetohydrodynamics. It is set in the framework of solar physics and astrophysics, but the focus is on basic concepts of plasma physics and on basic properties of plasmas. An average student in physics or mathematics at the K.U.Leuven does not know very much about plasma physics. He is not particularly worried by this lack of scientific knowledge as he happens to live in a corner of the universe where matter is predominantly solid, liquid, or gaseous. The three states of matter which occur at the surface of the earth are however not typical of matter in the universe. Most of the visible matter in the universe exists as plasma whereas lightning and the aurora are the only natural manifestations of the plasma state on earth. This Chapter is an exercise in public relations for plasma physics. Its aim is to show that plasmas are (almost) everywhere in the universe and to point out that they are extremely complicated physical systems fundamentally different from classic neutral gases, especially when there is a magnetic field present. The hope is that the reader is convinced that plasmas are exciting physical objects that are abundantly present in the universe and that he is prepared to make the effort to learn about the basic principles and properties of plasmas.

1.1 Plasma as the fourth state of matter

A plasma can be produced by raising the temperature of a gas until a reasonably high fractional ionization is obtained. A plasma is essentially a gas consisting of neutral and charged particles, ions and electrons, rather than of neutral atoms and molecules only as illustrated on Fig. 1.1. In general a plasma is **electrically neutral** overall, but the presence of charged particles means that a plasma can support electric currents and interact with electric and magnetic fields. It is important to be aware that a plasma cannot be treated simply as an ordinary gas which is electrically conducting. There is a fundamental difference between a neutral gas and a plasma that results from the different nature of the inter-particle forces. In a neutral gas the forces are very strong and short range, so that the dynamics of a neutral gas is dominated by two-body billiard-ball-like collisions. In a plasma the inter-particle forces between charged particles are electromagnetic forces. A charged particle interacts with the other charged particles through the Coulomb force. In addition a moving charged particle creates a magnetic field which produces a force on the other charged particles. The electromagnetic forces are comparatively weak and long-range. Due to the long range of the inter-particle forces each charged particle in a plasma interacts with a large number of other charged particles resulting in **collective plasma behaviour** : hence the fact that plasma is referred to as the fourth state of matter. *A plasma is a macroscopically electrically neutral substance containing many interacting free electrons and ions which exhibit collective behaviour due to the long-range Coulomb forces.*

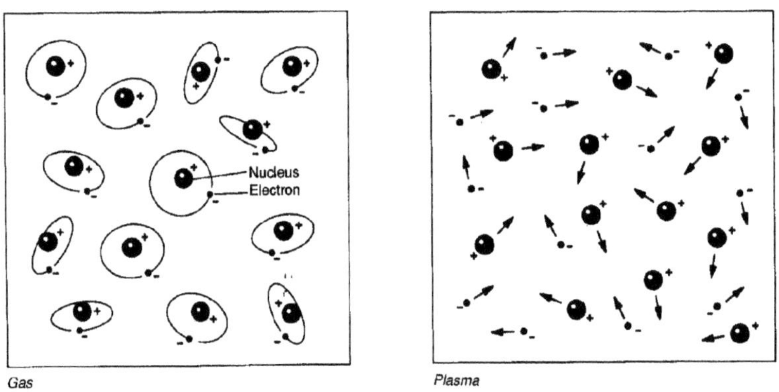

Figure 1.1: *Schematic representation of a gas and a plasma.*

1.2 Plasmas and magnetic fields

The interactions of a plasma with a magnetic field produce a wonderland of fascinating phenomena. In any room, you are largely unaware of the earth's magnetic field, because its interaction with the gas is negligible. If your room where filled with plasma, however, the situation would be very different. The magnetic field exerts a force on the plasma (Lorentz force) which can be split into two parts. The first part is a **magnetic pressure** which acts, just like ordinary gas pressure, from regions of high to low pressure. The second part is a **magnetic tension force** with the same effect as a tension in an elastic string. A curved magnetic field can eject plasma at a speed now known as the Alfvén speed or can support plasma against gravity. Since the magnetic tension provides a restoring force when a field line is curved, magnetic waves will propagate along the magnetic field lines in the same way as tension can make waves propagate along a string. Since a magnetic field exerts a force on a plasma, it may store energy. Plasma motions that twist and shear magnetic field lines can inject energy in the magnetic field. Occasionally, the magnetic field may become unstable and the stored energy be released in a violent disruption.

1.3 Why plasma physics

The desire to understand the basic properties of plasmas is quite recent. It is largely stimulated by the importance of plasma physics for solar physics, space physics, astrophysics, and for the development of controlled thermonuclear fusion. The ranges of temperature and number density of natural and man-made plasmas are huge as can be seen on Fig. 1.2 (borrowed from "Fusion - Physics of a fundamental energy source, Plasmas - the fourth state of matter, Characteristics of typical plasmas" at http://fusedweb.pppl.gov/CPEP/chart.html).

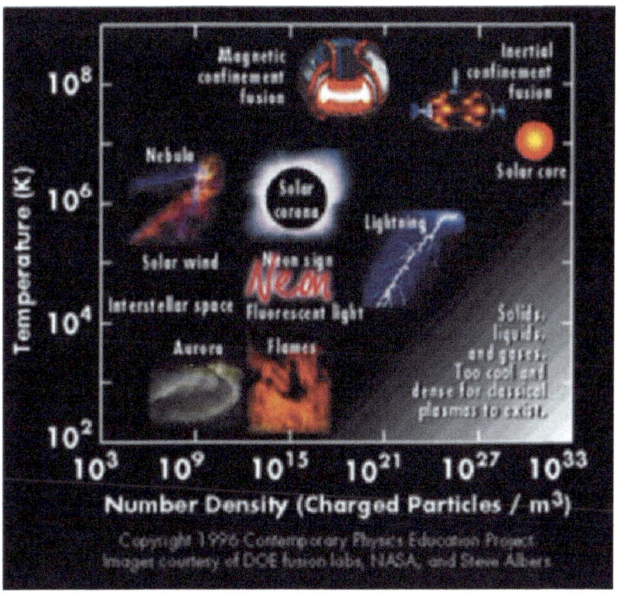

Figure 1.2: *Density and temperature of plasmas.*

The Sun and solar wind

The Sun can be regarded as a source of radiated energy which appears to have been relatively constant for periods of millions of years and not to have changed a lot in the past 4.6×10^9yr. The near constancy for the shorter times is easily understood once the origin of the radiated energy in nuclear reactions in the deep solar interior is appreciated. The Sun has such a large heat content that, if the central nuclear reactions were turned off, it would take 10^7yr before any knowledge of this reached the solar surface. The Sun appears to be a hot sphere of plasma held together by its own self-gravitation and kept hot by the steady nuclear reactions. This view of the Sun with slowly varying properties has led to the study of spherically symmetric models of the Sun, which evolve slowly in time, and with corresponding models for the structure and evolution of all types of stars. The structure and the different layers of the Sun are illustrated on Fig. 1.3.

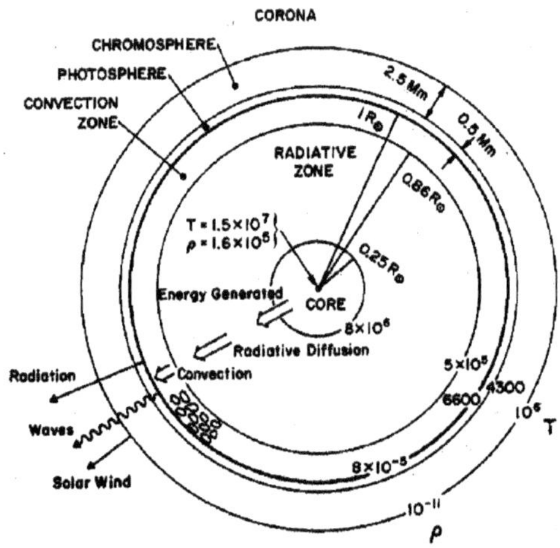

Figure 1.3: *Structure and different layers in the Sun.*

Superimposed on this very slowly varying Sun is an atmosphere of extreme complexity with rapid spatial and temporal variability which requires a totally different explanation. Only since the 1970s do we know that the solar atmosphere is controlled by the interactions of the magnetic field with the plasma. The old view of the solar atmosphere was a spherically symmetric structure with a completely unimportant magnetic field, except in sunspots. Elsewhere the magnetic field was thought to be uniform. Observations from satellites (e.g. Skylab, Solar Maximum Missions, Yohkoh, SOHO) and high resolution observations from the ground have overthrown this old view. We now know that over the whole of the solar *photosphere* the magnetic field is concentrated by turbulent convective motions into intense magnetic flux tubes of strength 1500 Gauss (a Gauss is a common unit of magnetic field strength in solar physics and astrophysics, 1 Gauss $= 10^{-4}$ Tesla) at the boundaries of convective elements. Going up above the photosphere of the Sun these flux tubes spread out in the *chromosphere* and fill the whole space in the upper chromosphere and the *corona* with beautiful magnetic loop structures where the magnetic field is strong enough. Elsewhere, the plasma stretches the magnetic field out into open structures leaving the Sun. Soft X-ray telescopes from space have revealed the corona in all its glory, emitting thermally at a few million K as can be seen on Fig. 1.4 (borrowed from "Yohkoh Public Outreach Project (YPOP)" at http://www.lmsal.com/YPOP/Spotlight/Tour/tour06.html). All the structure we see in the corona is caused by the magnetic field. If the Sun had no magnetic field, its atmosphere would be a rather dull object. There would not be any sunspots, intense flux tubes, prominences, spicules, flares, coronal mass ejections and the Sun would not have a

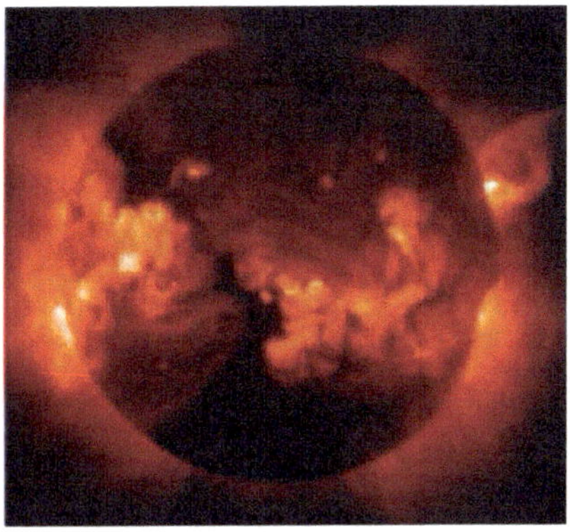

Figure 1.4: *Yohkoh image of the corona in soft X-rays, showing the bright magnetic loops.*

chromosphere, a corona or a wind. The magnetic activity of the Sun is not constant but changes with a period of about 11yr, known as the solar cycle. This is most clearly seen in the total number of sunspots which varies from a maximum to a minimum and back to a maximum in 11yr as is shown on Fig. 1.5 (borrowed from "Yohkoh Public Outreach Project (YPOP)" at http://www.lmsal.com/YPOP/Spotlight/Tour/tour07.html). The solar atmosphere has been identified as a gigantic plasma physics laboratory where the laws of modern plasma physics can be studied under conditions that cannot be realized on earth.

The Sun emits a highly conducting tenuous plasma, called the solar wind, at very high speeds into the interplanetary space. The existence of the solar wind is a consequence of the hot ($1 - 2 \times 10^6$ K) corona. It is not possible for a hot static corona to extend throughout interplanetary space. It must expand and as a result the Sun loses mass. The solar wind is far from being spherically symmetric. The high speed solar wind originates from the open magnetic field structures in the solar corona. The earth resembles a small pebble in a stream of plasma flowing supersonically out from the Sun. The expansion of the solar wind, combined with the solar rotation, has two consequences. First, the magnetic field, firmly rooted in the solar photosphere, is pulled outward since it is embedded in the radially outward-flowing plasma. Second, the magnetic field at larger distances is bent back azimuthally into a spiral as shown schematically on Fig. 1.6. The solar wind does not flow steadily. Solar activity manifests itself through the sunspot cycle; it causes the plasma from a particular solar region to expand at a much greater speed setting up a shock that accelerates ions in situ to large energies.

The magnetosphere

Plasma physics is also important closer at home. Although the main influence of the Sun on the Earth is through gravitation and electromagnetic radiation (primarily in the optical part of the spectrum), the Sun also interacts with the Earth through its particle emission in the solar wind. The reason why we do not have to worry about this high speed solar wind and the magnetic storms on the Sun is that we are lucky that the earth resides within a vast

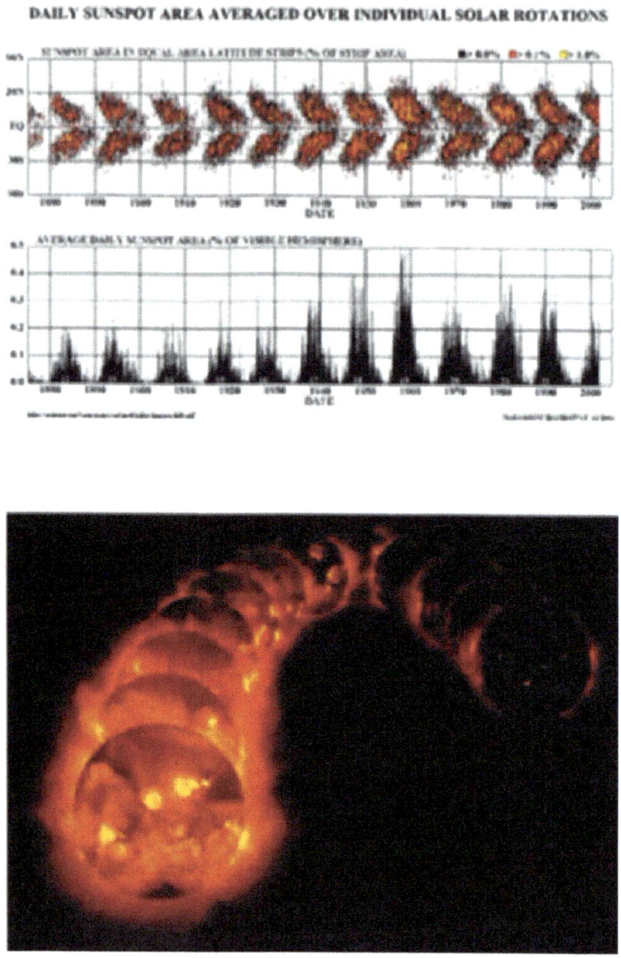

Figure 1.5: *Top: The number and position of the sunspots, represented in the so-called 'butterfly-diagram'; Bottom: 12 X-ray pictures taken between 1991 and 1995 showing the difference in solar radiative output when evolving from solar maximum to solar minimum.*

magnetic cavity, called the magnetosphere of the earth. The solar wind encloses the Earth and its local magnetic field in this magnetosphere as shown schematically on Fig. 1.7 (borrowed from "Oulu Space Physics Textbook" at http://www.oulu.fi/ spaceweb/textbook/). The solar wind reacts to the magnetic field of the earth and to the planetary magnetic fields as

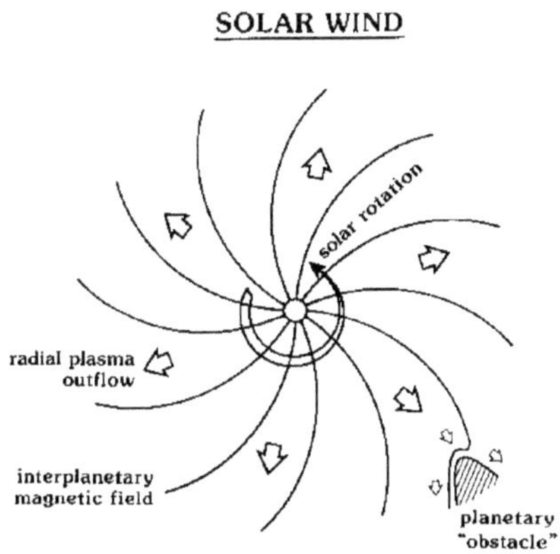

Figure 1.6: *Schematic representation of the spirals drawn by the magnetic field lines as they are advected outward by the solar wind. The wind itself flows radially outward from the Sun.*

obstacles in its path and is deflected by shock waves around the fields. These interactions create the planetary magnetospheres whose sizes depend on the strength of the magnetic field and the plasma pressure within the magnetospheres. The magnetosphere of the Earth, or of any other planet, is that region surrounding the planet in which its magnetic field has a controlling influence on, or dominates, the motions of energetic charged particles such as electrons, protons, or other ions.

In addition to changes induced by the rotation of the Sun, the energy transfer from the solar wind to the planetary magnetospheres varies with the solar magnetic cycle. At times of strong solar magnetic activity, the intensity of the solar wind increases and its interaction with the magnetosphere causes magnetic storms and aurorae. During solar maximum in 1989 geomagnetic disturbances and auroral displays could be observed as far south as Florida. Strong solar magnetic activity can cause a compression of the magnetosphere at its day-side to about half its size and an expansion to about twice its size at the night side. These changes in the size of the magnetosphere have an effect on artificial satellites due to increased drag and direct exposure to energetic particles in the solar wind. During the magnetic storms the magnetic field of the Earth is forced to change on a truly grand scale causing problems for geomagnetic navigation systems and disrupting radio communications. The disturbed magnetic fields can knock power plants out of service. During the severe geomagnetic storm on 13/03/1989 all of the Canadian province Quebec was plunged into complete darkness without warning and within a few seconds. So there is a practical interest in understanding the magnetosphere and its interaction with the solar wind. This has led to international research programmes in space weather. In addition the magnetosphere like the solar atmosphere is a plasma physics laboratory where we can observe and study plasmas under unique conditions that we cannot realize on Earth.

The heliosphere

The solar wind confines the planetary magnetic fields into magnetospheres but also produces

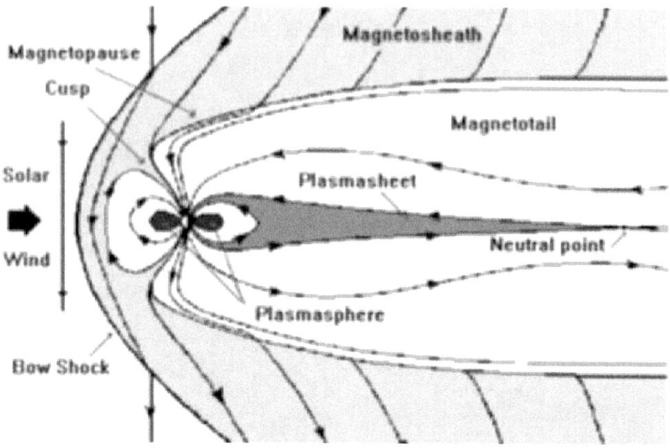

Figure 1.7: *The Earth's magnetosphere, deflected by the solar wind.*

a cavity surrounding the Sun in the local interstellar medium. The Sun has its own magnetosphere called the heliosphere; it is the region within the galactic medium where the solar plasma digs out, and fills a cavity. A hot corona cannot be in static equilibrium with the interstellar medium and must expand. Once the solar wind is introduced there is a surface surrounding the Sun at which the pressure of the solar wind balances the pressure of the interstellar gas and there is a heliospheric boundary whose structure is basically similar to that of the magnetosphere. The global solar magnetic field organizes the heliosphere. The morphology of the heliosphere, its evolution over space and time, and the location of its boundaries are determined by the global solar magnetic field and by the properties of the local interstellar medium. The position of the boundary is not known with great accuracy but is probably of order of 100 AU. The heliosphere contains most of the solar system but not the most distant comets.

Astrophysical plasmas

Almost all astrophysical objects are in the plasma state. Here we list a few examples of magnetic astrophysical plasmas. Obvious examples of stellar magnetic plasmas are the solar type stars. The only unique property of the Sun is its proximity. The Sun is our nearest star. It is so much nearer and looks so different from other stars that most people give the wrong answer to the question "what is the nearest star to us?" The second nearest known star to Earth is more than 2×10^5 times further away from the Earth than the Sun. Apart from being the star that is by far the closest to the Earth, the Sun is an ordinary main sequence (or dwarf) star of spectral type G2. This means that we can expect that many other stars, whose gross properties (mass, chemical composition, luminosity, effective temperature) are similar to those of the Sun, also exhibit atmospheric activity superimposed on their average behaviour. Because the Sun is so close to us, its activity is very apparent, but this would not be true even if the Sun were one of the other nearest stars to us. Because all other stars appear as point sources of radiation, it is impossible to observe their atmospheres with spatial resolution. Until the late 1970s the study of stellar activity relied on the existence of stars, whose activity is very much greater than that of the Sun. Ultraviolet spectroscopic observations from IUE, X-ray observations from the Einstein Observatory, microwave detections from the VLA and new optical observations from the ground have shown that solar-like activity (stellar spots, chromospheres, transition regions, coronae and stellar winds) indeed occurs in a wide variety of stars. If a solar-like star is defined as a star which has a turbulent magnetic field sufficiently strong to control the dynamics and energetics in its outer atmospheric regions, it then turns out that (i) dwarf stars of spectral type G-M and rapidly rotating subgiants and giants of spectral type F-K in spectroscopic binary systems are definitely solar-like, (ii) dwarf stars of spectral type A7-F7 are almost certainly solar-like, (iii) T Tauri stars and other pre-Main-Sequence stars are probably solar-like, (iv) slowly rotating single giants of spectral type F to early K are probably solar-like. As a consequence plasma physics is important for astrophysics as a whole.

The magnetic fields in the solar atmosphere are small scale magnetic fields concentrated in intense flux tubes and in sunspots in the photosphere. When averaged over the whole solar photosphere a small global field of a few Gauss is measured. The first discovery of global stellar magnetic fields goes back to the late 1940's when global magnetic fields were discovered in variable Ap stars. The Ap stars are peculiar stars with enhanced lines of the Fe-peak elements and greatly enhanced lines of the rare earth elements compared with the spectra of normal stars. The line strength anomalies are caused by atmospheric abundance anomalies confined in a thin layer in the atmosphere. The detected global magnetic fields vary in phase with the spectrum and light variations. The surface magnetic fields are predominantly dipolar, and their effective strengths range from a few hundred Gauss up to 34 kGauss. The periodic variations in spectrum, light, and magnetic field are explained with the oblique rotator model as due to the rotation of the star with the period of the observed variations equal to the period of rotation. The oblique rotator model assumes that the magnetic field is frozen in the stellar atmosphere and has an axis which is inclined to the axis of rotation, which itself is inclined to the line of sight. Because of the rotation of the star the observer sees different aspects of the dipolar magnetic field and measures a variable effective magnetic field. The spectrum and light variations are explained by assuming that the abundance anomalies are not uniform over the surface of the star. It is now clear that the magnetic Ap phenomenon (magnetic stars which are oblique rotators) extends from $T_e = 7400K$ to $T_e = 23000K$. Hence the name magnetic Ap is too narrow but is still used for historical reasons.

Far stronger global magnetic fields have been detected in white dwarfs and neutron stars. Of all the isolated white dwarfs that have been surveyed about 3% to 5% have observable magnetic fields with a strength from about 1×10^6 up to 5×10^8 Gauss. Very strong global magnetic fields are observed in neutron stars. Radio pulsars are a garden variety of neutron stars: hundreds have been detected since they were detected in 1968. They emit beams of radio waves which sweep through space as the stars rotate, like lighthouse beams, thus from afar pulsars seem to flicker or pulsate at their rotation period. Careful measurements have shown that pulsar periods increase over time, implying that the stars are gradually spinning down. This is attributed to their magnetic fields. The magnetic field is anchored to the neutron star surface, so as the star turns the field also must turn. This drives magnetic waves out, along with diffuse winds of charges particles (which emit the radio beams from just above the magnetic poles), carrying off energy and causing the star to slowly spin down. The magnetic fields that are computed from the observed periods and the observed variations of periods are for the majority of pulsars in the range $10^{11} - 10^{13}$ Gauss. The strongest magnetic fields ever detected occur in magnetars. Magnetars are highly magnetized neutron stars formed in a supernova explosion. The magnetic fields of magnetars are about 10^{15} Gauss. To put these enormous magnetic field strengths in perspective, let us compare them with that of more familiar objects. The Earth's magnetic field, which deflects compass needles, has a strength (measured at the N pole) of 0.6 Gauss; a common iron magnet, like those used to stick papers on a refrigerator, have a strength of 100 Gauss; the magnetic fields of sunspots have strengths of about 10^3 Gauss; the strongest man-made steady fields achieved in the laboratory have a strength of 4×10^5 Gauss; the maximum field observed on a ordinary stars is 10^6 Gauss and a typical magnetic field of radio pulsars has a strength of 10^{12} Gauss. White dwarfs and neutron stars require going beyond classic physics.

The interaction of magnetic fields with plasmas play an important role in star formation. The magnetic fields provide a very efficient mechanism of support for self-gravitating molecular clouds where star formation is taking place. They influence the conditions for collapse of a self gravitating cloud and help regulate star formation. Magnetic fields and plasma physics are also important for explaining the relatively slow rotation of non-degenerate stars. If during star formation angular momentum would be conserved on contraction, stars would be spinning much faster than they actually do. Without braking the Sun would not be a slow rotator with a period of 27 d but would rotate like mad with a rotational speed 10^8 times faster than its actual speed. Magnetic braking combined with a (magnetic) wind plays a fundamental role here. As a matter of fact all young Main sequence stars have undergone substantial braking during their formation, otherwise they would be spinning much faster than they actually do.

Controlled thermonuclear fusion

The Sun, like most stars, radiates an enormous amount of energy, because in its core the temperature and density are high enough to produce fusion of hydrogen into helium. The amount of hydrogen available for fusion is so large that the nuclear fusion reactor in the core can burn for about 5×10^9 years. Controlled nuclear fusion on earth would provide an almost unlimited and relatively clean means for energy production. The main obstacle in the way of harnessing this source of energy is the fact that the reactions will take place at a useful rate only if the temperature of the material is of the order of 10^8 K. Material at this temperature is ionized. A way to confine and control this very hot plasma is by using magnetic fields in toroidal devices. Early attempts to do this revealed that a magnetic plasma is a far more complicated system than had been anticipated. This has triggered off a programme of theoretical and experimental research into the properties of magnetic plasmas which still continues.

1.4 Aim of the course

The goal of the course is not to cover the various subjects of the previous Section. That would be far too ambitious and actually a full course can be dedicated to each of these subjects. However, before we can embark on a study of any of these subjects, we need to learn about the fundamental concepts of plasma physics and the basic properties of plasmas. Hence, the aim is to give a basic introduction to plasma physics with an emphasis on Magnetohydrodynamics (MHD for short). Since this course is the first encounter with plasma physics for students at the K.U.Leuven, its level is elementary. Since both students in mathematics and physics can take this course, both mathematical integrity and physical intuition are pursued.

The content might not be what is routinely expected from an introductory course on plasma physics. Indeed, there are different options about what to teach in such an introductory course. A possible approach is to present a selection of plasma physics phenomena to convince the reader that plasma physics is indeed an important subject worthwhile pursuing and then to introduce the appropriate approximate plasma model to explain the phenomenon under discussion. Even if the selection is narrowed down to e.g. the Earth and our Sun, this would be a huge task if a reasonable level of depth is pursued and would require a variety of approximate plasma models if the observations are to be underpinned by theory. Of course, time and space can be gained by presenting the plasma models with a minimal discussion of their theoretical foundations. However, this obscures the overall logical structure of theoretical plasma physics.

A plasma is a large N-body system of mobile charged particles and electromagnetic fields. A complete simulation of such a macroscopic system by integration of the Lorentz force equations and the corresponding microscopic Maxwell equations is far beyond our reach, even with the most powerful computers. Even if we could solve the system exactly, we would have far more information than we would require. For these reasons a number of plasma models have been developed. The models range from kinetic models which contain all the relevant physical phenomena, but are still largely unsolvable, to fluid models which selectively remove small-scale physics, but are more tractable and yield very useful large-scale solutions.

Since it is impossible to cover all of the plasma models in a first course, I focus on Magnetohydrodynamics (MHD for short). MHD is a macroscopic, non-relativistic theory that is concerned with large-scale (global) and low-frequency (slow) phenomena in magnetic plasmas. This focus reflects a personal bias as I have been using MHD to a large extent, but not exclusively, as a mathematical model for carrying out my research on magnetic plasmas. However, I have convinced myself that there are indeed good reasons for starting with MHD. Firstly, MHD can be viewed as an extension of classic hydrodynamics. It can be expected that students in their third year are familiar with very basic results of classic fluids. Hence, it makes sense to introduce to these students a plasma model that they can link to a classic fluid, on their first encounter with plasmas. Secondly, MHD gives an accurate description of many of the complicated interactions of magnetic fields with the plasmas of the Sun and stars and of fusion machines. Magnetohydrodynamics can be used to study the global equilibrium, stability, waves and heating of the basic magnetic structures in the solar atmosphere and of plasmas in thermonuclear magnetic fusion machines. Thirdly, the model of ideal MHD has an attractive mathematical structure with well-defined conservation laws.

MHD can be viewed as classical fluid dynamics with the additional complication that the fluid is electrically conducting. A possible way of introducing MHD is to write down the constitutive equations of classical fluid dynamics and to add the terms and equations due to the magnetic field. This approach might be the best way to economically present the equations of MHD and it allows us to move on to various applications of MHD without much ado. However, this leaves the student without any idea how and where to place MHD in the wider context of plasma physics. I have opted to put an emphasis on fundamental concepts and first principles. The goal of this introductory course is to acquaint the student with basic properties of magnetic plasmas. Therefore, I shall begin from first principles with the fundamental microscopic equations and then systematically derive the equations of multi-fluid and single fluid MHD. I start from the Boltzmann equation and derive the equations of MHD as moment equations of the Boltzmann equation. This approach is longer and more tedious, but has the considerable advantage it allows to show where and how microphysics has been removed and to determine the limitations and the domain of validity of MHD. The student will not end up with the wrong impression that MHD covers all of plasma physics. He/she will be fully aware that plasmas can show behaviour that cannot be captured with MHD. In that sense the course is also an invitation to the interested reader to go beyond classic MHD and discover many fascinating plasma behaviour that is not dealt with here.

> Facts are of not much use, considered as facts. They bewilder by their number and apparent incoherency. Let them be digested in theory, however, and brought into mutual harmony, and it is another matter. Theory is the essence of facts. Without theory scientific knowledge would only be worthy of the mad house.
> *Electromagnetic theory*
> O. Heaviside.

I decided to set this course in the framework of solar and space plasma physics and astrophysics. Again this decision reflects a personal bias as my research is in solar physics and plasma astrophysics. Again, I am convinced that this is a good choice. The fully and partially ionized plasmas that are the central focus of solar and space physics are related on a fundamental level to laboratory plasma physics, which directly investigates basic plasma physical processes, and to astrophysics, a discipline that relies heavily on understanding the physics of the plasma state. Although solar physics is the framework in which I want to set my course, I have refrained myself from concentrating on a description of plasma physics phenomena in solar physics and astrophysics. The reader should by now understand that the emphasis of this course is on basic concepts of plasma physics and on basic properties of plasmas. If he/she expects a course that mainly concentrates on a description of plasma physics phenomena in solar physics and astrophysics then he/she will be disappointed. Observational facts are essential in sciences. However, in order to fully appreciate the observations of the complicated behaviour of plasmas, the reader should have learned first the basic concepts.

This course gives a well-structured presentation of basic concepts and fundamental principles of plasma physics and MHD and paves the way to a wide variety of subjects where plasma physics plays an important role. Of course, once we have a well-defined mathematical theory for describing a plasma system, we need to show its strength by applying it to specific situations. Why would we bother with the effort of setting up a mathematical model, if it were not for explaining observations and experiments. The fact that the reader sees that MHD does work, is a strong motivation for going through its sometimes tedious derivations. The fifth Chapter is dedicated to MHD waves in uniform and unbounded plasmas. This can be seen as a first elementary application of MHD and/or theory on very low-frequency waves in magnetic plasmas. In the sixth and last Chapter we use hydrodynamics and MHD to study the solar wind. It is the only Chapter that deals exclusively with solar physics.

The amount of mathematics and physics required for this course is limited. A knowledge of vector calculus, real calculus and electromagnetic theory are the modest prerequisites from mathematics and physics. I like to think that both students in applied mathematics and physics benefit from a course where by starting from basic principles and by using realistic approximations a mathematical model is constructed for a complicated physical system as a magnetic plasma. This mathematical model contains all the information of the system under study within the limitations of the approximations. It can be used to explain observed behaviour and to predict future behaviour of the system. In a sense this course can be seen as an example of mathematical modelling of complicated systems. Unfortunately, the mathematical model equations are often not co-operative and it often requires hard work to extract information from them. The mathematics involved in this course is most of the time rather elementary. I am not worried about this elementary level of mathematics and physics. This is an introductory course with the emphasis on basic concepts. In the last Chapter the mathematical model is applied to the solar wind. Here the mathematics is a little bit more advanced as we are required to solve non-linear ordinary differential equations. The theoretical analysis remains pretty straightforward, but the actual computation of the solutions turns out to be a tough problem of numerical mathematics.

The exercises take a special place in this course. We all learn best the things that we have discovered for ourselves. Hence, in stead of being very detailed in the derivation of the equations and results, I have taken the relaxed attitude to be economical on intermediate results and steps when these are straightforward. A first class of exercises invites the reader to fill in gaps in the often long derivation of equations. There is no need for the reader to memorize the different steps required for obtaining a given equation or result, but he/she should have gone through this straightforward mathematics at least once. Also, it is difficult, if not impossible, for anyone to learn a subject purely by reading about it. Applying the information to specific problems and thereby being encouraged to think about what has been read, is essential in the learning process. A second class of exercises tries to invite the reader to just do that. The exercises form a major part of this book. In the Belgian educational system students do have to take exams. The exam for the material covered in this book consists of solving the exercises and explaining the solutions by using the notes the students have prepared themselves.

References

> What one man can invent another can discover.
> *Sherlock Holmes, The Adventure of the Dancing Men*
> Sir Arthur Conan Doyle

The reader is advised to consult the following good textbooks in order to get some perception of what the present course has tried to teach him of the vast field of plasma physics and MHD. The books on plasma physics and magnetohydrodynamics that I find very helpful and inspiring are

- W. Baumjohann and R. A. Treumann, " Basic Space Plasma Physics", Imperial College Press, 1997. (BT1997).
- J.A. Bittencourt, " Fundamentals of Plasma Physics", Pergamon Press, 1986. (B1986).
- R.O. Dendy, "Plasma Dynamics", Oxford Science Publications, 1990. (D1990).
- J.P. Freidberg, " Ideal Magnetohydrodynamics", Plenum Press, 1987. (F1987).
- J.P. Goedbloed, "Lecture Notes on Ideal Magnetohydrodynamics", Rijnhuizen Report 83-145, 1983. (G1983).

- R.J. Goldston and P.H. Rutherford, " Introduction to Plasma Physics", Institute of Physics Publishing, 2000. (GR2000).
- R.D. Hazeltime and F. L. Waelbroeck, " The Framework of Plasma Physics", Frontiers in Physics, Perseus Books, 1998. (HW1998).
- R.V. Polovin and V.P. Demutskii, " Fundamentals of Magnetohydrodynamics", translated from Russian by D. ter Haar, Consultants Bureau, New York and London, 1990. (PD1990).
- P.H. Roberts, "An Introduction to Magnetohydrodynamics", Longmans, 1967. (R1967).
- G. L. Siscoe, "Solar System Magnetohydrodynamics", in "Solar-Terrestrial Physics, Principles and Theoretical Foundations", edited by R.L. Carovillano and J.M. Forbes, D. Reidel Publishing Company, 1983. (S1983).
- P.A. Sturrock, "Plasma Physics, An Introduction to the Theory of Astrophysical, Geophysical and Laboratory Plasmas", Cambridge University Press, 1994. (S1994).
- L.C. Woods, "Principles of Magnetoplasma Dynamics", Clarendon Press, Oxford, 1987. (W1987).

The books on solar magnetohydrodynamics that I frequently use are

- A.W. Hood, "The Sun: An Introduction to MHD", http://www-solar.mcs.st-andrews.ac.uk/, 2000. (H2000).
- E. R. Priest, "Solar Magnetohydrodynamics", D. Reidel Publishing Company, 1984. (P1984).
- E. R. Priest, "Solar System Magnetic Fields" , D. Reidel Publishing Company, 1985. (P1985).
- M. Stix, "The Sun", Springer-Verlag, 1989. (S1989).

When teaching and writing the chapter on the solar wind I was happy to consult

- J.C. Brandt, "Introduction to the Solar Wind", Freeman, 1970. (B1970).
- P. Charnonneau, "Large Scale Dynamics of the Solar Wind", unpublished class notes, 1994. (C1994).
- A.J. Hundhausen, "Coronal Expansion and Solar Wind", Springer-Verlag, 1972. (H1972).

When writing the section on collisions I found the book

- T.W.B. Kibble and F.H. Berkshire, "Classical Mechanics",Longman, 1996. (KB1996)

very helpful. Very useful information on physical constants can be found in the little book

- J.D. Huba "Physical constants, NRL Plasma Formulary", Naval Research Laboratory, 2000.

Chapter 2

Basic plasma properties

> Like all other arts, the Science of Deduction and Analysis is one which can only be acquired by long and patient study, nor is life long enough to allow any mortal to attain the highest possible perfection in it. Before turning to those moral and mental aspects of the matter which presents the greatest difficulties, let the enquirer begin by mastering more elementary problems.
> *Sherlock Holmes, A study in Scarlet*
> Sir Arthur Conan Doyle

A plasma is an ionized gas that is in a state of electrical quasi-neutrality, the behaviour of which is governed by collective effects due to the long range electromagnetic interaction between the charged particles. In this Chapter we shall study the two basic characteristics that are used in the definition of a plasma : quasi-neutrality and collective behaviour. We shall determine basic consequences arising from the long-range Coulomb interactions and we shall point out the necessity of taking into account the collective behaviour of many charged particles brought about by the long-range interactions. Plasma oscillations and Debye screening are typical examples of this collective behaviour; the plasma thus strongly exhibits a medium-like behaviour. We identify major plasma parameters that characterize the high-frequency behaviour associated with the dynamics of the electrons and the low-frequency behaviour associated with the dynamics of the ions.

2.1 Elements of plasma kinetic theory

Particle distribution functions

A plasma is a system containing a very large number of mobile charged particles. Each charged particle creates its own microscopic electric and magnetic fields and reacts to the microscopic fields of all other particles. The actual electric and magnetic fields are the sum of all the microscopic contributions of the particles. These fields have an extremely complicated spatial structure and vary on different time scales. A complete simulation of a plasma by integration of the Lorentz force equations for all the particles and the corresponding microscopic Maxwell equations is far beyond our reach, even with the most powerful computers. Solving the classic electromagnetic many-body problem for a plasma is a hopeless and hardly interesting task. It is hardly interesting because it would give us far more information than required. We are not interested in knowing the position and velocity of each individual particle at any given time.

Hence we *replace the real plasma consisting of discrete particles with a smeared-out density distribution function in phase space.* This might be expected to be reasonable if each particle feels the effect of many other particles simultaneously and not just that of a few of its nearest neighbours. This is what happens when there are many particles in the Debye sphere, which we shall discuss in Sections 2.3 - 2.5. In analogy with the *configuration space* defined by the position coordinates x, y, z, it is convenient to consider the *phase space* defined by the six coordinates x, y, z, w_x, w_y, w_z.

An element of volume in configuration space is represented by $\Delta^3\vec{r} = \Delta x\ \Delta y\ \Delta z$. This is a finite element volume, sufficiently large to contain a large number of particles, yet sufficiently small in comparison with the characteristic lengths associated with the spatial variation of physical quantities as, for example, density and temperature. When we refer to a particle as being situated inside $\Delta^3 r$, at position $\vec{r}$, it is meant that its x coordinate lies between x and $x + \Delta x$, its y coordinate between y and $y + \Delta y$, and its z coordinate between z and $z + \Delta z$. Particles localized in $\Delta^3\vec{r}$, at position $\vec{r}$, may have completely arbitrary velocities which are represented by scattered points in velocity space.

An element of volume in velocity space is represented by $\Delta^3\vec{w} = \Delta w_x\ \Delta w_y\ \Delta w_z$. For a particle to be included in $\Delta^3\vec{w}$, around the terminal point of the velocity vector $\vec{w}$, its x component of velocity must lie between w_x and $w_x + \Delta w_x$, its y component of velocity between w_y and $w_y + \Delta w_y$, and its z component between w_z and $w_z + \Delta w_z$.

In phase space an element of volume can be imagined as the volume of a six-dimensional cube: $\Delta^3\vec{r}\Delta^3\vec{w} = \Delta x\ \Delta y\ \Delta z\ \Delta w_x\ \Delta w_y\ \Delta w_z$. Note that inside $\Delta^3\vec{r}\Delta^3\vec{w}$, at the position $(\vec{r}, \vec{w})$ in phase space, there are only the particles inside $\Delta^3\vec{r}$ around $\vec{r}$ whose velocities lie inside $\Delta^3\vec{w}$ around $\vec{w}$. The number of points inside a volume element $\Delta^3\vec{r}\Delta^3\vec{w}$ is, in general, function of time t and of the position in the phase space. The coordinates $\vec{r}$ and $\vec{w}$ of phase space are independent variables, since they label individual volume elements in phase space. In particular, $\vec{w}$ is not the velocity of an individual particle, and it is not the fluid velocity of a plasma element.

The *distribution function* in phase space, $f_\alpha(\vec{r}, \vec{w}, t)$ is the density of the representative points in phase space for particles of type α, that is, the number of particles of type α $\Delta^6 N_\alpha(\vec{r}, \vec{w}, t)$ inside $\Delta^3\vec{r}\Delta^3\vec{w}$, at the position $(\vec{r}, \vec{w})$ is

$$\Delta^6 N_\alpha(\vec{r}, \vec{w}, t) = f_\alpha(\vec{r}, \vec{w}, t) \times \Delta^3\vec{r}\Delta^3\vec{w} \tag{2.1}$$

The distribution function does not depend on the coordinates of all the single particles of a given species; it only depends on the phase space coordinates $(\vec{r}, \vec{w})$ and time t. The exact positions of the particles have been smeared out over the phase space volume $\Delta^3\vec{r}\Delta^3\vec{w}$ and the distribution function does not describe the exact positions of the particles in this volume. It is assumed that the density of the representative points in phase space does not vary rapidly from one element of volume to its neighbouring element, so that $f_\alpha(\vec{r}, \vec{w}, t)$ can be considered a continuous function of its arguments. $f_\alpha(\vec{r}, \vec{w}, t)$ is also a positive and finite function at any time. Also in an element of volume with very large velocity coordinates (w_x, w_y, w_z), the number of representative points has to be relatively small, since in any macroscopic system, there can only be relatively few particles with very large velocities. In particular $f(\vec{r}, \vec{w}, t)$ must tend to zero as the velocity becomes infinitely large. The types of particles that occur in a plasmas are electrons, ions and also neutrals when the plasma is only partially ionized.

The distribution function is, in general, a function of the position vector $\vec{r}$. When this is the case it is said to be *nonuniform*. When it is independent of position, the distribution function is *uniform*. In velocity space the distribution function can be *anisotropic*, when it depends on the orientation of the velocity vector $\vec{w}$, or *isotropic*, when it does not depend on the orientation of $\vec{w}$, but only on its magnitude. The description of different plasmas requires the use of uniform and nonuniform, isotropic and anisotropic, time independent and time dependent distribution functions.

Macroscopic quantities

The distribution functions contain all the information on the system under study. Once we know $f_\alpha(\vec{r}, \vec{w}, t)$, we can compute all macroscopic fluid quantities related to the particles of type α and to the plasma as a whole. The particle density in configuration space of particles of species α, $n_\alpha(\vec{r}, t)$, is the integral of the particle distribution function over velocity space. The mass density in configuration space of particles of species α, $\rho_\alpha(\vec{r}, t)$ and the mass density of the plasma as a whole are then readily computed:

$$
\begin{aligned}
n_\alpha(\vec{r}, t) &= \int f_\alpha(\vec{r}, \vec{w}, t) d^3\vec{w} \\
\rho_\alpha(\vec{r}, t) &= n_\alpha(\vec{r}, t) m_\alpha \\
n(\vec{r}, t) &= \sum_\alpha n_\alpha(\vec{r}, t) \\
\rho(\vec{r}, t) &= \sum_\alpha \rho_\alpha(\vec{r}, t)
\end{aligned}
\tag{2.2}
$$

m_α is the mass of a particle of species α. The velocity of the particles of type α as a whole, $\vec{v}_\alpha(\vec{r}, t)$, is the weighted mean of $\vec{w}$ with f_α as weighting function; the bulk velocity of the plasma $\vec{v}(\vec{r}, t)$ is the weighted mean of the velocities $\vec{v}_\alpha(\vec{r}, t)$ of the different species with the mass densities $\rho_\alpha(\vec{r}, t)$ as weighting factors:

$$
\begin{aligned}
\vec{v}_\alpha(\vec{r}, t) &= <\vec{w}> = \frac{1}{n_\alpha(\vec{r}, t)} \int \vec{w} f_\alpha(\vec{r}, \vec{w}, t) d^3\vec{w} \\
\rho\vec{v}(\vec{r}, t) &= \sum_\alpha \rho_\alpha \vec{v}_\alpha(\vec{r}, t)
\end{aligned}
\tag{2.3}
$$

Once the fluid velocity of the particles of species α is determined, we can define the random velocities

$$
\vec{u}_\alpha(\vec{r}, t) = \vec{w} - \vec{v}_\alpha(\vec{r}, t) \tag{2.4}
$$

of the particles of type α with respect to fluid of the particles of type α. We do not expect the system to have reached thermal equilibrium, nevertheless we use the mean kinetic energy of these random motions $\frac{m_\alpha}{2} <\mid \vec{u}_\alpha(\vec{r},t) \mid^2>$ to define the thermal velocity $v_{t,\alpha}$ and the temperature T_α of the particles of type α. The temperature T of the plasma as a whole is a measure of the kinetic energy of the random motions with respect to the whole plasma:

$$v_{t,\alpha}^2 = \frac{1}{3} <\mid \vec{u}_\alpha(\vec{r},t) \mid^2> = \frac{1}{n_\alpha} \int \mid \vec{u}_\alpha(\vec{r},t) \mid^2 f_\alpha(\vec{r},\vec{w},t) d^3\vec{w} = \frac{k_B T_\alpha}{m_\alpha}$$

$$3k_B n T = \sum_\alpha n_\alpha m_\alpha <\mid \vec{w} - \vec{v}(\vec{r},t) \mid^2> = \sum_\alpha m_\alpha \int \mid \vec{w} - \vec{v}(\vec{r},t) \mid^2 f_\alpha(\vec{r},\vec{w},t) d^3\vec{w} \quad (2.5)$$

Here k_B is the Boltzmann constant $k_B = 1.3807 \times 10^{-23}$J K^{-1}. This temperature is the *kinetic temperature*, a quantity which we can formally calculate for any type of distribution function. Therefore it is not necessarily a true temperature in the thermodynamic sense, which can only be calculated for plasmas in or close to thermal equilibrium. This kinetic temperature is rather a measure for the spread of the particle distribution in velocity space. Moreover, because each particle species may have its own distribution function, the kinetic temperatures of the plasma components may differ from each other. In addition, in an anisotropic plasma the temperatures parallel and perpendicular to the magnetic field are in general different, because the particle distributions have different dependencies in the parallel and perpendicular directions. The thermal velocity $v_{t,\alpha}$ defined in (2.5) is the "root-mean-square" of the random velocities in any one direction. Sometimes its square is defined without the factor 1/3 in (2.5). The average kinetic energy of a particle of type α due to random motions with respect to the fluid of particles α is

$$\frac{m_\alpha}{2} <\mid \vec{u}_\alpha(\vec{r},t) \mid^2> = \frac{3m_\alpha}{2} v_{t,\alpha}^2 = \frac{3}{2} k_B T_\alpha$$

There is not necessarily equipartition of kinetic energy of random motions in the three spatial directions. When this equipartition is absent, it makes sense to define thermal velocities and temperatures in more than one direction.

The electric charge density Q_α and the electric current density $\vec{j}_\alpha$ for particles of type α are computed with the number density $n_\alpha(\vec{r},t)$, the electric charge per particle q_α and velocity of the particles of type α as a whole. The total electric charge density Q and the total electric current density $\vec{j}$ are obtained by summing Q_α and $\vec{j}_\alpha$ over the different species α:

$$Q_\alpha(\vec{r},t) = n_\alpha(\vec{r},t) q_\alpha, \quad \vec{j}_\alpha(\vec{r},t) = Q_\alpha(\vec{r},t)\vec{v}_\alpha(\vec{r},t)$$

$$Q(\vec{r},t) = \sum_\alpha Q_\alpha(\vec{r},t), \quad \vec{j}(\vec{r},t) = \sum_\alpha \vec{j}_\alpha(\vec{r},t) \quad (2.6)$$

Let us recall that the real plasma consisting of discrete particles is replaced with a smeared-out density distribution function in phase space. The fields involved in this description are global smeared out fields $\vec{E}$ and $\vec{B}$ associated with the total electric charge density Q and the total electric current density $\vec{j}$. They satisfy the global macroscopic Maxwell equations.

Maxwell's equations

The equations of Maxwell are well known

$$
\begin{aligned}
\nabla \times \vec{E} &= -\frac{\partial \vec{B}}{\partial t} \\
\nabla . \vec{B} &= 0 \\
\nabla \times \vec{B} &= \mu \vec{j} + \frac{1}{c^2}\frac{\partial \vec{E}}{\partial t} \\
\nabla . \vec{E} &= \frac{Q}{\epsilon}
\end{aligned}
\qquad (2.7)
$$

In these equations $\vec{E}$ is the electric field, $\vec{B}$ is the magnetic induction, $\vec{j}$ is the total electric current density, Q is the total electric charge density and μ and ϵ are the magnetic permeability and the electric permittivity of the plasma. The latter two quantities are almost always replaced by their values for vacuum $\mu_0 = 4\pi \times 10^{-7}$H m^{-1} and $\epsilon_0 = 8.8542 \times 10^{-12}$F m^{-1} so that $1/(\mu_0\epsilon_0) = c^2$ with $c = 2.9979 \times 10^8$m s^{-1} the velocity of light. The magnetic field $\vec{H}$ and the electrical displacement $\vec{D}$ have been eliminated from Maxwell's equations (2.7) by the use of

$$\vec{H} = \frac{\vec{B}}{\mu_0} \quad \wedge \quad \vec{D} = \epsilon_0 \vec{E}$$

Since $\vec{H}$ will not be used in what follows, we shall refer to $\vec{B}$ as the magnetic field.

The Boltzmann equation

The evolution of the distribution function $f_\alpha(\vec{r}, \vec{w}, t)$ in space and time is governed by a partial differential equation that is known as the *Boltzmann equation*. We are looking at particles of type α with mass m_α and electrical charge q_α. Recall that $\Delta^6 N_\alpha(\vec{r}, \vec{w}, t) = f_\alpha(\vec{r}, \vec{w}, t)\Delta^3\vec{r}\Delta^3\vec{w}$ is the number of particles of species α which are contained in the element of volume $\Delta^3\vec{r}\Delta^3\vec{w}$ in phase space at the position $(\vec{r}, \vec{w})$ and at time t. Let the particles be subjected to an *external force* $\vec{F}_\alpha(\vec{r}, \vec{w}, t)$. For example, for particles in an electromagnetic field and a gravitational field, this force is

$$\vec{F} = m_\alpha \vec{g} + q_\alpha(\vec{E} + \vec{w} \times \vec{B})$$

where $\vec{g}$ is the gravitational acceleration.

In absence of particle interactions, all particles in the volume element $\Delta^3\vec{r}\Delta^3\vec{w}$ in phase space at the position $(\vec{r}, \vec{w})$ and at time t will be found in the volume $\Delta^3\vec{r'}\Delta^3\vec{w'}$ in phase space at the position $(\vec{r'}, \vec{w'})$ and at time $t' = t + \Delta t$ with

$$\vec{r'} = \vec{r} + \vec{w}\,\Delta t, \quad \vec{w'} = \vec{w} + \vec{a}\,\Delta t, \quad \vec{a} = \frac{\vec{F}_\alpha}{m_\alpha}$$

$\vec{a}$ is the particle acceleration. For what follows it is instructive to note that Δt is a short time interval since what we really have in mind is to take the limit $\Delta t \to 0$!

The number of particles in the element of volume $\Delta^3\vec{r'}\Delta^3\vec{w'}$ at the position $(\vec{r'}, \vec{w'})$ and at time $t' = t + \Delta t$ is

$$\Delta^6 N_\alpha(\vec{r'}, \vec{w'}, t') = f_\alpha(\vec{r'}, \vec{w'}, t')\Delta^3\vec{r'}\Delta^3\vec{w'}$$

Let us now write $\Delta^6 N_\alpha(\vec{r'}, \vec{w'}, t')$ in terms of the old variables $(\vec{r}, \vec{w})$ and $\Delta^6 N_\alpha(\vec{r}, \vec{w}, t)$. For the volume elements we have the simple result

$$\Delta^3\vec{r'}\Delta^3\vec{w'} = J\Delta^3\vec{r}\Delta^3\vec{w}$$

where $J = \det M$ is the Jacobian of the transformation from $(\vec{r}, \vec{w})$ to $(\vec{r'}, \vec{w'})$.

$$M = \frac{D(\vec{r'}, \vec{w'})}{D(\vec{r}, \vec{w})}$$

is a 6×6 matrix. It is left as an exercise to the reader to determine the elements of the matrix M. The relevant observation to make is that all elements on the diagonal of M are equal to 1 and that all off-diagonal elements either vanish or are proportional to Δt. Hence $J = 1 + O((\Delta t)^2)$.

It is a basic assumption that $f_\alpha(\vec{r}, \vec{w}, t)$ is a continuous and differentiable function. We use a Taylor expansion to obtain

$$\begin{aligned} f_\alpha(\vec{r'}, \vec{w'}, t') &= f_\alpha(\vec{r} + \vec{w}\ \Delta t, \vec{w} + \vec{a}\ \Delta t, t + \Delta t) \\ &= f_\alpha(\vec{r}, \vec{w}, t) + \vec{w}.\nabla_x f_\alpha(\vec{r}, \vec{w}, t)\ \Delta t + \vec{a}.\nabla_w f_\alpha(\vec{r}, \vec{w}, t)\ \Delta t + \frac{\partial f_\alpha}{\partial t}\ \Delta t + O((\Delta t)^2) \end{aligned}$$

where ∇_x and ∇_w denote the $\nabla-$operator in configuration space and in velocity space respectively. Hence

$$\Delta^6 N_\alpha(\vec{r'}, \vec{w'}, t') - \Delta^6 N_\alpha(\vec{r}, \vec{w}, t) =$$

$$\left\{ \left(\frac{\partial f_\alpha}{\partial t} + \vec{w}.\nabla_x f_\alpha(\vec{r}, \vec{w}, t) + \vec{a}.\nabla_w f_\alpha(\vec{r}, \vec{w}, t) \right) \Delta t + O((\Delta t)^2) \right\} \Delta^3\vec{r}\Delta^3\vec{w}$$

In absence of collisions the number of particles remains unchanged and the right hand side of the previous equation is zero. When we divide this equation by the volume element $\Delta^3\vec{r}\Delta^3\vec{w}$ and also by Δt and then take the limit $\Delta t \to 0$ we obtain *the collisionless Boltzmann equation* or *Vlasov equation*

$$\frac{\partial f_\alpha}{\partial t} + \vec{w}.\nabla_x f_\alpha(\vec{r}, \vec{w}, t) + \vec{a}.\nabla_w f_\alpha(\vec{r}, \vec{w}, t) = 0 \qquad (2.8)$$

However when collisions are present $\Delta^6 N_\alpha(\vec{r'}, \vec{w'}, t') - \Delta^6 N_\alpha(\vec{r}, \vec{w}, t) \neq 0$ since some of the particles which were initially in the volume element $\Delta^3\vec{r}\Delta^3\vec{w}$ may be removed from this volume element and particles which were initially out of the volume element $\Delta^3\vec{r}\Delta^3\vec{w}$ may end up in the volume element $\Delta^3\vec{r'}\Delta^3\vec{w'}$ The net variation (gain or loss) of particles of type α per unit volume in phase space and unit time due to collisions is denoted as

$$\left(\frac{\partial f_\alpha}{\partial t}\right)_{\text{coll}}.$$

The difference in the number of particles is then

$$\Delta^6 N_\alpha(\vec{r'}, \vec{w'}, t') - \Delta^6 N_\alpha(\vec{r}, \vec{w}, t) = \left(\frac{\partial f_\alpha}{\partial t}\right)_{\text{coll}} \Delta^3\vec{r}\Delta^3\vec{w}\Delta t$$

With the same simple mathematical manipulations that we used to obtain the collisionless Boltzmann equation, we now find the classic *Boltzmann equation:*

$$\frac{\partial f_\alpha}{\partial t} + \vec{w}.\nabla_x f_\alpha(\vec{r}, \vec{w}, t) + \vec{a}.\nabla_w f_\alpha(\vec{r}, \vec{w}, t) = \left(\frac{\partial f_\alpha}{\partial t}\right)_{\text{coll}} \tag{2.9}$$

This equation was first derived by Boltzmann in 1872. Recall that in the Boltzmann description there are two types of forces that act on the particles. First there are the long-range Lorentz force $q_\alpha(\vec{E} + \vec{w} \times \vec{B})$, and the gravitational force $m_\alpha \vec{g}$ with $\vec{E}$, $\vec{B}$ and $\vec{g}$ global fields. Second there are the collisions which are related to short-range forces on the length scale of the Debye sphere. For plasmas of astrophysical and fusion interest, the dominant collisions are elastic Coulomb collisions. The Boltzmann equation (2.9) supplemented with an equation for the collision term and the Maxwell equations (2.7) form a closed set of equations. Once $f_\alpha(\vec{r}, \vec{w}, t)$ is known, any macroscopic quantity, such as density, pressure, temperature, can be computed by integration of expressions involving $f_\alpha(\vec{r}, \vec{w}, t)$ over velocity space.

Examples of distribution functions

Let us now look at the *equilibrium state of a system of particles that is free from external forces.* In the equilibrium state the particle interactions do not cause any change in the distribution function with time and there are no spatial variations in the particle number density. The equilibrium distribution function is given by the *Maxwell-Boltzmann distribution function*:

$$f(w_x, w_y, w_z) = n\left(\frac{m}{2\pi k_B T}\right)^{3/2} \exp\left(-\frac{m(w_x^2 + w_y^2 + w_z^2)}{2k_B T}\right) \tag{2.10}$$

The number density n and the temperature T are constants. The Maxwell-Boltzman distribution function is time independent, uniform and isotropic. Whatever the velocity distribution of a system of particles initially not in equilibrium may be, it tends to the Maxwell-Boltzmann distribution (2.10) in the course of time, if the system is maintained isolated from the action of external forces.

The distribution function $g(w_x)$ for the velocity component w_x, can be obtained by integrating over w_y and w_z

$$g(w_x) = n\left(\frac{m}{2\pi k_B T}\right)^{1/2} \exp\left(-\frac{m w_x^2}{2k_B T}\right) \tag{2.11}$$

A similar expression applies to w_y and w_z. Each of the velocity components has a *Gaussian* distribution, with average value $< w_i >= 0, \quad i = x, y, z$.

The distribution function $g(w_x)$ and also the Maxwell-Boltzmann function (2.10) $f(w_x, w_y, w_z)$ are properly normalized as $\int_{-\infty}^{+\infty} g(w_x) dw_x = n, \quad \int f(w_x, w_y, w_z) d\vec{w} = n$ and n is indeed the number density of the particles in configuration space.

The fact that $< w_x >, < w_y >, < w_z >$ are zero, means that there is no translational motion of the system as a whole. The velocities $\vec{w}$ are completely random velocities. On the other hand, $< w_i^2 > \quad i = x, y, z$ is intrinsically positive since $\frac{m}{2} < w^2 >= \frac{3m}{2} < w_x^2 >$ is the mean kinetic energy of the random motions of the particles. The Maxwell-Boltzmann function (2.10) has anticipated the fact that the mean kinetic energy of the random motions is related to the temperature. A simple calculation shows that

$$v_t^2 = \frac{< w^2 >}{3} =< w_x^2 >= \frac{1}{n}\int_{-\infty}^{+\infty} w_x^2 g(w_x) dw_x = \frac{k_B T}{m}$$

The average kinetic energy of a particle due to random motions is

$$\frac{m}{2} < w^2 >= \frac{3m}{2} v_t^2 = \frac{3}{2} k_B T$$

and there is equipartition of energy in the 3 spatial directions. The thermal velocity can be used to rewrite the Maxwell-Boltzmann function (2.10) as

$$f(w_x, w_y, w_z) = \frac{n}{(\sqrt{2\pi} v_t)^3} \exp\left(-\frac{w^2}{2v_t^2}\right) \tag{2.12}$$

In view of what follows let us recall that (i) number density n and the temperature T are constants, (ii) the Maxwell-Boltzmann distribution function (2.10) is time independent, uniform and isotropic and (iii) the particles have only kinetic energy and have no potential energy associated with their position.

The classic Maxwell-Boltzmann distribution (2.10) has $< \vec{w} >= 0$ so that there is no macroscopic velocity present in the system. A macroscopic velocity $\vec{v} = (v_x, v_y, v_z)^t$ can introduced by using a *"shifted" Maxwell-Boltzmann distribution function* as

$$f(w_x, w_y, w_z) = n \left(\frac{m}{2\pi k_B T}\right)^{3/2} \exp\left(-\frac{m \mid \vec{w} - \vec{v} \mid^2}{2k_B T}\right) \tag{2.13}$$

It is straightforward to show that $< w_x >= v_x$, $< w_y >= v_y$, $< w_z >= v_z$, $< \vec{w} >= \vec{v}$ so that $\vec{w} - \vec{v} = \vec{u}$ is the random velocity of the particles. Here also it has been anticipated that the mean kinetic energy of the random motions is related to the temperature, since

$$v_t^2 = \frac{< u^2 >}{3} =< u_x^2 >= \frac{k_B T}{m}$$

The average kinetic energy of a particle due to random motions is

$$\frac{m}{2} < u^2 >= \frac{3m}{2} v_t^2 = \frac{3}{2} k_B T$$

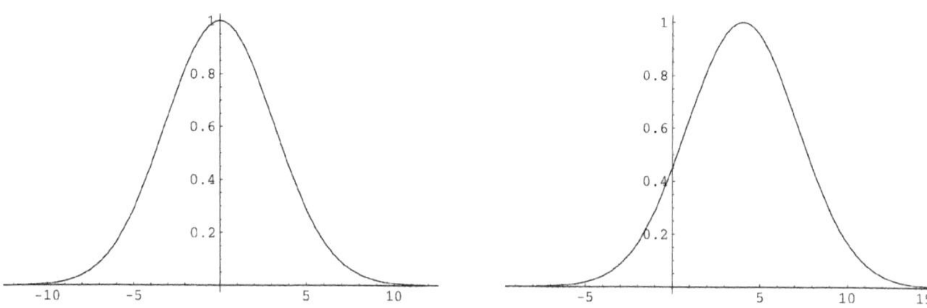

Figure 2.1: *Maxwellian and shifted Maxwellian distribution functions.*

In some cases, a plasma has an anisotropic distribution function, which can be approximated as *a "bi-Maxwellian" with a different temperature along the magnetic field than across the magnetic field.* In this case, taking the direction of the magnetic field to be the z direction, we have

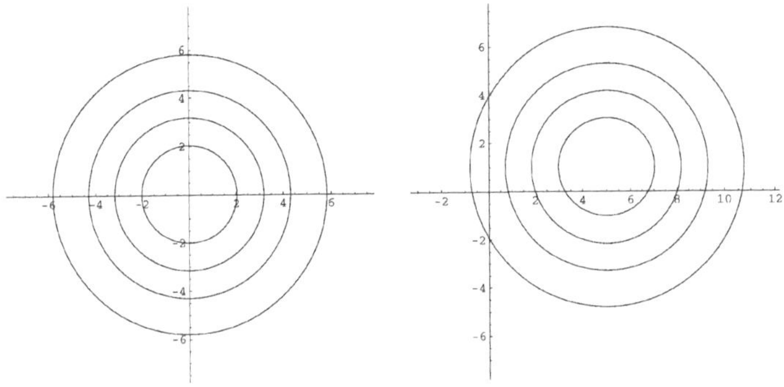

Figure 2.2: *Maxwellian and shifted Maxwellian distribution functions.*

$$f(w_{||}, w_{\perp}) = \frac{n}{(\sqrt{2\pi} v_{t||})(\sqrt{2\pi} v_{t\perp})^2} \exp\left(-\frac{w_{\perp}^2}{2v_{t\perp}^2} - \frac{w_{||}^2}{2v_{t||}^2}\right) \tag{2.14}$$

Here

$$v_{t||}^2 = \frac{k_B T_{||}}{m}, \quad v_{t\perp}^2 = \frac{k_B T_{\perp}}{m}$$

$w_{||}$ is the component of $\vec{w}$ parallel to $\vec{B}$ and $w_{\perp}^2 = w_x^2 + w_y^2$.

The graphic representation of $g(w_x)$ for a Maxwellian and a shifted Maxwellian are given in Fig. 2.1. The isolines of $f(w_{||}, w_{\perp})$ for a Maxwellian, and a shifted Maxwellian are given in Fig. 2.2. The isolines of $f(w_{||}, w_{\perp})$ for a bi-Maxwellian are given in Fig. 2.3; the right hand part of this figure shows a bi-Maxwellian with a beam of fast particles.

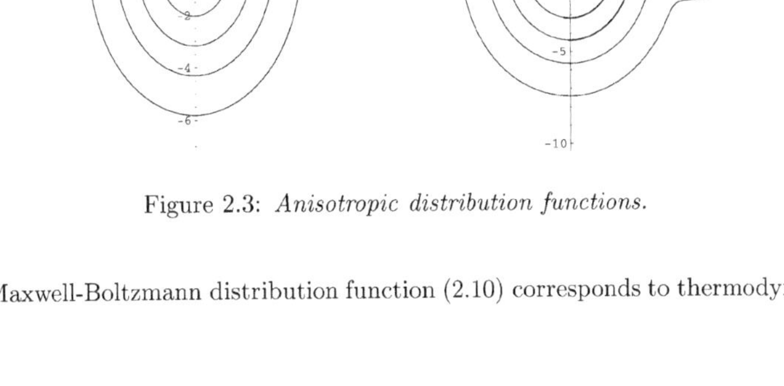

Figure 2.3: *Anisotropic distribution functions.*

The Maxwell-Boltzmann distribution function (2.10) corresponds to thermodynamic equilibrium and uniform distributions of density and temperature. This is in sharp contrast with the ubiquitous presence of spatial and temporal variations in density, pressure and temperature in nature. The assumption of full thermodynamic equilibrium has to be relaxed. Often we are dealing with a system of particles that, although not in equilibrium, is not very far from it. It is then a good approximation to assume that, in the neighbourhood (element of volume) of any point in the system, there is an equilibrium situation described by a local Maxwell-Boltzmann distribution function. For larger spatial scales gradients of T may exist, but these scales are so large that, locally, the equilibrium is not perturbed. In addition, time variations of T may exist, but on such a slow time scale that instantaneous equilibrium is a very good approximation. The assumption of *full thermodynamic equilibrium* is replaced with *local thermodynamic equilibrium (LTE)*. The *isotropic Maxwell-Boltzmann distribution function for nonuniform, time dependent LTE* is

$$f(\vec{r},\vec{w},t) = n(\vec{r},t)\left(\frac{m}{2\pi k_B T(\vec{r},t)}\right)^{3/2} \exp\left(-\frac{m \mid \vec{w}-\vec{v}(\vec{r},t)\mid^2}{2k_B T(\vec{r},t)}\right) \tag{2.15}$$

The generalization to an anisotropic Maxwell-Boltzmann distribution function for nonuniform, time dependent LTE is straightforward.

In a plasma we have at least two types of particles, electrons and ions. Hence we have at least two thermal velocities $v_{t,e}$ and $v_{t,i}$ defined as

$$v_{t,e}^2 = \frac{k_B T_e}{m_e}, \quad v_{t,i}^2 = \frac{k_B T_i}{m_i} \tag{2.16}$$

m_e and m_i are the mass of an electron and of an ion respectively, for a proton $m_i = m_p$. Note that $m_e = 9.1094 \times 10^{-31}$kg and $m_p = 1.6726 \times 10^{-27}$kg, so that $m_p/m_e = 1.8336 \times 10^3$ and $(m_p/m_e)^{1/2} = 42.81$. The ratio of the thermal velocities $v_{t,e}$ and $v_{t,i}$ is then

$$\frac{v_{t,e}}{v_{t,i}} = \left(\frac{T_e}{T_i}\right)^{1/2}\left(\frac{m_i}{m_e}\right)^{1/2}$$

In case that the electrons and ions have comparable temperatures, i.e. $T_e \approx T_i$ then

$$\frac{v_{t,e}}{v_{t,i}} = \left(\frac{m_i}{m_e}\right)^{1/2} >> 1 \tag{2.17}$$

2.2 Plasma oscillations : the plasma frequency

In the absence of external disturbances a plasma is macroscopically neutral. This means that under equilibrium conditions with no external forces present, in a volume sufficiently large to contain a large number of particles and yet sufficiently small compared with the characteristic lengths for variation of macroscopic quantities such as density and temperature, the net resulting electric charge is zero. In the interior of the plasma the microscopic space charge fields cancel each other and no net space charge exist over a macroscopic region.

Let us now see what electrical quasi-neutrality means by looking at what happens when there are deviations from charge neutrality in the plasma. In this Section we are interested in the high-frequency oscillations that occur in a fluid consisting of electrons and ions when the particles are displaced relative to one another. The electrons and the ions are treated as interpenetrating fluids. We take the economic principle of minimal effort for maximal result as guideline and we use a very simple mathematical model for describing the plasma. We hope that the relevant physics is contained in this mathematical model and that the physics that we leave out is unimportant. We assume that there are not any random thermal motions. This idealized system is referred to as a *cold plasma*. The particle distribution function of a cold plasma is a δ-function centered on the average velocity.

Here I am going to commit a major pedagogical crime. I am going to use the two-fluid equations for a plasma consisting of an electron and ion fluid although I shall derive these equations in the following Chapter. I know this is bad behaviour, but the other option is to start from the equation of motion for a single electron and to use intuitive geometric arguments. I do not like the latter option as it hides the fact that the phenomenon under study is a fluid phenomenon. The two-fluid equations that will be used here are simplified variants of the general two-fluid equations that will be obtained in the following Chapter. The electric force is the only force that we allow to act on the electrons and ions. Since the plasma is cold there is no pressure force and we forget about the magnetic force. The relevant equations are the equation of conservation of mass and the equation of motion for the electron and ion fluids:

$$\frac{\partial n_e}{\partial t} + \nabla.(n_e\vec{v}_e) = 0$$

$$\rho_e\left(\frac{\partial}{\partial t} + \vec{v}_e.\nabla\right)\vec{v}_e - Q_e\vec{E} = 0$$

$$\frac{\partial n_i}{\partial t} + \nabla.(n_i\vec{v}_i) = 0$$

$$\rho_i\left(\frac{\partial}{\partial t} + \vec{v}_i.\nabla\right)\vec{v}_i - Q_i\vec{E} = 0$$

$$\varepsilon_0\nabla.\vec{E} = Q_i + Q_e, \quad Q_e = -n_e e, \quad Q_i = n_i e \tag{2.18}$$

We start with a plasma where the electron and ion fluids are motionless and where the electrons and ions are initially uniformly distributed, so that the plasma is electrically neutral everywhere. The equations for this uniform static background are:

$$\vec{v}_{e,0} = 0, \quad \vec{v}_{i,0} = 0$$

$$n_{e,0} = n_{i,0} = \text{constant}$$

$$Q_{e,0} = -n_{e,0}e = -n_{i,0}e = -Q_{i,0}$$

$$\vec{E}_0 = 0 \tag{2.19}$$

The subscript 0 refers to equilibrium quantities. This system is now perturbed by displacing the electron and the ion fluids from their equilibrium positions. Any physical quantity $f(\vec{x}, t)$, for instance the particle density of the electrons n_e, is now dependent on space and time and is written as the sum of its constant equilibrium part f_0 and a perturbed part $f_1(\vec{x},t)$:

$$f(\vec{x},t) = f_0 + f_1(\vec{x},t)$$

$$n_e(\vec{x},t) = n_{e,0} + n_{e,1}(\vec{x},t) \tag{2.20}$$

Since the ions are much heavier than the electrons ($m_p/m_e = 1.8336 \times 10^3$), we start by treating the heavy ions as immobile in comparison to the light electrons and set

$$\vec{v}_{i,1} = 0, \quad n_{i,1} = 0, \quad Q_{i,1} = 0$$

We now introduce the decomposition (2.20) into the original time and space dependent equations (2.18). The assumption is that the perturbed quantities $f_1(\vec{x},t)$ are small compared to the equilibrium quantities f_0 so that when rewriting equations (2.18) it suffices to only retain linear terms in the perturbed quantities. This leads to the following set of linear equations

$$\frac{\partial n_{e,1}}{\partial t} + n_{e,0}\nabla.\vec{v}_{e,1} = 0$$

$$m_e\frac{\partial \vec{v}_{e,1}}{\partial t} = -e\vec{E}_1$$

$$\varepsilon_0\nabla.\vec{E}_1 = -en_{e,1} \tag{2.21}$$

It is now straightforward to derive an equation for the perturbation of the number density of the electrons $n_{e,1}$

$$\left\{\frac{\partial^2}{\partial t^2} + \frac{e^2 n_{e,0}}{m_e\epsilon_0}\right\} n_{e,1} = 0 \tag{2.22}$$

This is the standard equation of a linear oscillator. The coefficient of $n_{e,1}$ must have the dimension of frequency squared. The solution of this equation is found by taking $n_{e,1} \sim \exp(i\omega t)$, where ω is the angular frequency of the oscillation which is equal to $\omega_{p,e}$, the electron plasma frequency:

$$\boxed{\omega_{p,e}^2 = \frac{n_{e,o}e^2}{m_e\epsilon_0} \qquad (2.23)}$$

The electron plasma frequency is a fundamental plasma quantity characteristic for the dynamics of the electrons. To obtain quantitative information, we replace the universal constants $e = 1.6022 \times 10^{-19}$C, $m_e = 9.1094 \times 10^{-31}$kg, and $\epsilon_0 = 8.8542 \times 10^{-22}$F m^{-1} by their numerical values and we write n_0 as (n_0 / characteristic density) $\times$ characteristic density. We choose the characteristic density to be that of a plasma in a present-day medium-size fusion experiment, 10^{19}m^{-3}. The plasma electron frequency can then be written in the form

$$\omega_{p,e} = 1.8 \times 10^{11}\left(\frac{n_e}{10^{19}\text{m}^{-3}}\right)^{1/2}\text{rad} \times \text{s}^{-1}$$

Thus, when $n_e = 10^{19}\text{m}^{-3}$, we have $\omega_{p,e} = 1.8 \times 10^{11}\text{rad} \times \text{s}^{-1}$. This makes it clear that we are dealing with high-frequency oscillations.

In retrospect, this plasma oscillation could have been predicted on the argument that the transfer of electrons from a given region of space to a neighbouring region induces a charge separation. This local charge gives rise to an electric field E. Since the electrons are much lighter than the ions, they respond much more rapidly to the electric field, and the motion of the ions can be neglected in first instance. The electric field pulls the electrons back to their initial position in order to reduce the local charge separation which is the source of the electric field. Since the electron possesses a finite mass, those electrons that begin to move cannot stop at the exact state of equilibrium, they overshoot the target and produce another non-equilibrium distribution in the opposite direction. The electrons begin to move in the reverse direction, overshoot the equilibrium again, and so on. The electrons perform an oscillatory motion, with the Coulomb force acting as the restoring force and the mass of the electron as the inertia. This is called a plasma oscillation. This oscillatory process could in principle continue indefinitely, as energy shifts from the electrostatic field to kinetic electron energy and back again, for ever. The fact that the electron mass, though small, is non-zero, is essential. If the electron had no mass, the electrostatic energy could not be transformed into electron kinetic energy. The small electron mass accounts for the rapid response of the electrons to the electric field, and the high value of $\omega_{p,e}$, which is typically in the microwave range of frequencies for present-day fusion plasmas.

Recall that we have treated the ions as immobile. We now allow the ions to move:

$$\vec{v}_{i,1} \neq 0, \quad n_{i,1} \neq 0, \quad Q_{i,1} \neq 0$$

The relevant equations are

$$\begin{aligned} \frac{\partial n_{e,1}}{\partial t} + n_{e,0}\nabla.\vec{v}_{e,1} &= 0 \\ m_e \frac{\partial \vec{v}_{e,1}}{\partial t} &= -e\vec{E}_1 \\ \frac{\partial n_{i,1}}{\partial t} + n_{i,0}\nabla.\vec{v}_{i,1} &= 0 \\ m_i \frac{\partial \vec{v}_{i,1}}{\partial t} &= +e\vec{E}_1 \\ \varepsilon_0 \nabla.\vec{E}_1 &= e(n_{i,1} - n_{e,1}) \end{aligned} \tag{2.24}$$

As before, it is now straightforward to derive an equation for the perturbation of the number density of the electrons $n_{e,1}$

$$\left\{ \frac{\partial^2}{\partial t^2} + \frac{e^2 n_{e,0}}{m_e \epsilon_0} \frac{m_i + m_e}{m_i} \right\} n_{e,1} = 0 \tag{2.25}$$

Again we find a linear oscillator. The angular frequency ω of the oscillation is $\omega = \omega_p$

$$\omega_{p,i}^2 = \frac{n_{i,0}e^2}{m_i\epsilon_0}$$

$$\omega_p^2 = \frac{e^2 n_{e,0}}{m_e\epsilon_0}\frac{m_i + m_e}{m_i} = \omega_{p,e}^2 + \omega_{p,i}^2 \qquad (2.26)$$

$\omega_{p,i}$ is the ion plasma frequency. Of course

$$\frac{\omega_{p,e}}{\omega_{p,i}} = \left(\frac{m_i}{m_e}\right)^{1/2} >> 1 \qquad (2.27)$$

and thus

$$\omega_p \approx \omega_{p,e}$$

The plasma electron frequency is the characteristic frequency with which a fluid consisting of electrons and ions oscillates when the particles are displaced relative to one another. It is a quantity that is related to the high-frequency dynamics of the electrons. The very light electrons react to space charge separation and try to restore electrical neutrality. The massive ions are outrun by the light electrons and do not participate in restoring electrical neutrality.

Let us now look at the effect of the plasma oscillation on the electrical quasi-neutrality of the plasma. The plasma oscillation is a periodic oscillation of the electric charge and constitutes a periodic violation of charge neutrality. The obvious conclusion is that charge neutrality does not hold when viewed over time spans τ that are comparable to the period of the plasma oscillation: $2\pi/\omega_{p,e}$ which is 3.5×10^{-11}s for a plasma with $n_0 = 10^{19}\text{m}^{-3}$. This does not worry us too much since the average of the oscillating electric charge over one period of the plasma oscillation is of course 0. This means that *electric neutrality is a good approximation of reality* for time spans τ sufficiently longer than the plasma oscillation period:

$$\tau \gg 2\pi/\omega_{p,e} \qquad (2.28)$$

Now wait a minute!!! This Section begun by saying that the net resulting charge is zero in a volume sufficiently large to contain a large number of particles. Here we are talking about sufficiently long time spans. We shall come back to this result in Section 4 of the present Chapter.

The plasma oscillation defines a natural microscopic time unit which we can use to calibrate what we mean by global macroscopic time spans.

The plasma oscillations have been studied by using the cold plasma model in which the random thermal motions of the electrons and the ions were ignored. These oscillations are therefore a completely ordered response of the plasma to electrostatic perturbations.

2.3 The Debye shielding length

We now take up the problem of determining the effective interaction between charged particles in a plasma. We shall calculate an effective potential field around a point positive charge (this will be our Coulomb scattering centre in Section 7) by taking explicit account of the statistical distribution of other charged particles. This calculation will lead to the notion of *Debye screening.* The potential field around a charged particle is effectively screened by the cloud of the other charged particles; its force range is now confined within a certain characteristic length, called the Debye shielding length, determined by the density and the temperature of the plasma. Let us return to the initial unperturbed cold plasma introduced in the previous Section, and insert into it a point positive test charge, q_i. We know from our first year course on electromagnetism that for an isolated positive charge, in free space, the electric field is directed radially outward and given by

$$\vec{E} = \frac{1}{4\pi\epsilon_0}\frac{q_i}{r^2}\vec{1}_r$$

so that the electrostatic Coulomb potential $\phi_C(r)$ due to this isolated charged particle in free space, is

$$\phi_C(r) = \frac{q_i}{4\pi\epsilon_0 r} \qquad (2.29)$$

with r the distance to the positive test charge and $\vec{1}_r$ the unit vector in the radial direction. In the plasma the spatial distribution of charged particles is affected by the presence of such a potential field and deviates from a uniform distribution. The charge attracts a cloud of electrons and repels the local ions, so that it is completely shielded from the rest of the plasma. Outside the cloud, there is no electric field.

The mathematical model of a cold plasma introduced in the previous Section does not suffice any longer. It misses an essential ingredient for the present discussion, we need thermal motions. Life gets more complicated but it will be more fun. Let us now introduce the random thermal motions and raise the temperature of the electrons and ions from absolute zero to T. For present purposes, we assume that *the plasma is isothermal*, that is to say at a constant temperature, independent of position.

Deep inside the cloud, the thermal motions of the electrons will not be sufficient to enable them to escape from the vicinity of the point charge q_i. At the edge of the cloud, most of the charge q_i is screened by the inner electrons. In this outer region, the thermal energy of the electrons may exceed their electrostatic potential energy in the field of the largely screened charge q_i. Such electrons are able to escape from the cloud; to the extent that this occurs , q_i is no longer so effectively screened. We can find a self-consistent solution for the electrostatic potential $\phi(r)$, which arises from the charge q_i, and the response of the plasma to the presence of q_i. A calculation along these lines was originally carried out by Debye and Huckel in 1923 in connection with the theory of screening in a strong electrolyte.

A single particle has now both kinetic energy and potential energy:

$$\frac{mw^2}{2} + q_{e,i}\phi(r)$$

and the *isotropic, time independent distribution function* is now

$$f(\vec{r},\vec{w}) = n_0\left(\frac{m}{2\pi k_B T}\right)^{3/2}\exp\left(-\frac{mw^2/2 + q_{e,i}\phi(r)}{k_B T}\right)$$

Here $q_e = -e$ for electrons and $q_i = Ze$ for ions. For simplicity we assume that electrons and ions have the same temperature $T_e = T_i = T$ and we take $Z = 1$.

This distribution is nonuniform. The non-uniformity, that is the dependence on position, is hidden in $\phi(r)$. As a consequence the particle (electron / ion) number density depends on position, the factor n_0 will turn out to be the particle density at infinity. We obtain the number density of the electrons, n_e, and that of the ions, n_i, by integration of the distribution function over velocity space:

$$n_{e,i}(r) = \int f(\vec{r}, \vec{w}) d^3\vec{w} = n_o \exp\left(-\frac{q_{e,i}\phi(r)}{k_B T}\right)$$

For $Z = 1$ the charge density Q, including the test charge $q_i = e$, is

$$Q(r) = e(n_i - n_e) + e\delta(\vec{r}) = -2n_0 e \sinh(\frac{e\phi(r)}{k_B T}) + e\delta(\vec{r})$$

Poisson's equation for the potential $\phi(r) : \nabla^2\phi = -Q/\epsilon_0$ takes the form

$$\nabla^2\phi = \frac{2n_0 e}{\epsilon_0}\sinh(\frac{e\phi}{k_B T}) - \frac{e}{\epsilon_0}\delta(\vec{r})$$

This is a difficult equation to solve, the main reason being that sinh is a transcendent (and non-linear) function of its argument. We know the disease. The mathematical cure is to replace sinh by its linear Maclaurin polynomial. This is a mathematically sound operation if the absolute value of the argument of sinh is much smaller than 1,

$$\frac{e\phi}{k_B T} \ll 1$$

but it remains to be seen whether operation is also physically sound. The inequality means that the electrostatic energy $e\phi$ associated with q_i is much less than the thermal energy $k_B T$. We shall come back to this important assumption. We can then approximate

$$\sinh(\frac{e\phi}{k_B T}) \approx \frac{e\phi}{k_B T}$$

so that Poisson's equation for ϕ becomes

$$\nabla^2\phi = \frac{2}{\lambda_D^2}\phi - \frac{e}{\epsilon_0}\delta(\vec{r})$$

The parameter

$$\lambda_D = \left(\frac{\epsilon_0 k_B T}{n_0 e^2}\right)^{1/2} = \left(\frac{k_B T}{m_e}\right)^{1/2}\frac{1}{\omega_{p,e}} = \frac{v_{t,e}}{\omega_{p,e}} \qquad (2.30)$$

introduced here has the dimension of a length and is called *the Debye length*; it is an important parameter for the study of plasmas. Choosing typical values appropriate to present-day medium-size fusion experiments, with characteristic thermal energy 1 keV and number density $10^{19}m^{-3}$, we can rewrite the expression for λ_D as

$$\lambda_D = 7.4 \times 10^{-5}\left(\frac{(k_B T)/(1\text{keV})}{n_e/(10^{19}\text{m}^{-3})}\right)^{1/2} \text{m}$$

For a plasma at 1 keV with $n_0 = 10^{19}\text{m}^{-3}$, for example, λ_D takes the value 7.4×10^{-5}m.

The problem has spherical symmetry since all quantities depend only on the distance r to the test charge. In spherical coordinates we can rewrite Poisson's equation (for $r \neq 0$) as

$$\frac{1}{r^2}\frac{d}{dr}\left(r^2\frac{d\phi(r)}{dr}\right) = \frac{2}{\lambda_D^2}\phi$$

In order to solve this equation we use our conservative reflex and have a look at the solution for a positive point test charge in vacuum. Since close to the test charge the electrostatic potential $\phi(r)$ should be the same as the electrostatic Coulomb potential $\phi_C(r)$ we seek a solution of the form

$$\phi(r) = \phi_C(r)y(r)$$

where $y(0) = 1$ of course. Substitute this expression for $\phi(r)$ into Poisson's equation and find the following differential equation for $y(r)$:

$$\frac{d^2y}{dr^2} = \frac{2}{\lambda_D^2}y(r)$$

The solution to this equation that is finite for $r \to +\infty$ is

$$y(r) = \exp(-\sqrt{2}\frac{r}{\lambda_D})$$

Hence, the solution to Poisson's equation with the boundary condition that $\phi(r)$ vanishes at infinity is

$$\phi(r) = \frac{q_i}{4\pi\epsilon_0 r}\exp(-\sqrt{2}\frac{r}{\lambda_D}) \qquad (2.31)$$

The physical meaning of this result for $\phi(r)$ is clear. The Debye length (2.30) is a measure for the range of the effect of the test charge. As can be seen on Fig. (2.4) the effective potential is essentially equivalent to the bare Coulomb potential (2.29) for distances much smaller than λ_D. However, for distances larger than λ_D the effective potential decreases exponentially. The potential around a point charge is effectively screened out by the induced space-charge field in the plasma for distances greater than the Debye length. The Debye length and the range of the effect of the test charge q_i are larger in a hot diffuse plasma than in a cool dense plasma. This is to be expected : if T is high, more electrons in the cloud at a given distance will be able to escape, so that q is less efficiently screened; if n_0 is small, electrons will have to be drawn from a larger volume in order to shield a given charge q. The sphere that has radius λ_D and is centered on q_i is known as the Debye sphere of q_i. Its volume is $V_D = \frac{4\pi}{3}\lambda_D^3$.

2.4 Charge neutrality again

Let us now look at what has happened to charge neutrality and compute the total charge in a sphere that has radius L and is centered on q_i. Recall that the total charge density $Q(r)$ is $Q(r) = e(n_i - n_e) + e\delta(r)$. The total charge is

$$TC = \int_0^L e(n_i - n_e)4\pi r^2 dr + e$$

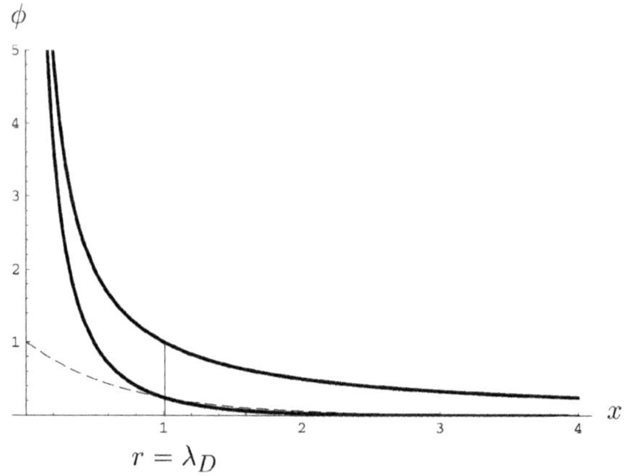

Figure 2.4: *The classic Coulomb potential for a point source and its Debye shielded modification. The dashed line is the graph of the exponential function with which the Coulomb potential is multiplied in order to obtain the Debye shielded potential.*

We again approximate sinh by the linear term of its Maclaurin expansion so that

$$n_i - n_e \approx -2n_0 \frac{e\phi(r)}{k_B T} = \frac{-2\sqrt{2}\exp(-x)}{4\pi\lambda_D^3} \frac{}{x}$$

The dimensionless spatial coordinate x, is defined as $x = \sqrt{2}r/\lambda_D$. Integration then gives the amazingly simple result:

$$TC = e(1 + L_*)\exp(-L_*) \qquad (2.32)$$

First of all note that

$$\lim_{L_* \to +\infty} TC = 0$$

The charge of the test particle is neutralized by the charge distribution surrounding the test particle. This neutralization effectively takes place over a distance of a few Debye lengths. In order to make that clear let us evaluate TC for 4 values of L namely

$$L = \lambda_D, \quad 3\lambda_D, \quad 5\lambda_D, \quad 10\lambda_D$$

which correspond to

$$L_* = \sqrt{2}, \quad 3\sqrt{2}, \quad 5\sqrt{2}, \quad 10\sqrt{2}$$

to find that

$$TC \approx 0.59e, \quad \approx 0.075e, \quad \approx 0.0069e, \quad \approx 4 \times 10^{-6}e$$

Within a distance of $10\lambda_D$ the charge neutralization is effectively completed. In the close proximity of the test charge there is an imbalance of charge and as a consequence an electric field. The total charge is negligibly small if $L > 10\lambda_D$. Overall quasi-neutrality is a good approximation of reality for any volume with a length scale L much larger than λ_D:

$$L \gg \lambda_D \qquad (2.33)$$

A collection of particles behaves as an overall quasi-neutral substance only if the preceding inequality holds for its dimension L. This inequality is *the first criterion for the definition of a plasma.* If the dimension L is near or less than λ_D, then our approximations break down and significant deviations from quasi-neutrality occur.

Here we have a condition for the length scales over which quasi-neutrality is realized. How is this condition related to the condition on the time scales (2.28) ? As a matter of fact these conditions are equivalent. Recall that $\lambda_D = v_{t,e}/\omega_{p,e}$ to rewrite the first criterion (2.33) as

$$L/v_{t,e} \gg 1/\omega_{p,e}$$

Now

$$\frac{L}{v_t} = \tau_t$$

is the time it takes a thermal signal with velocity v_t to cross the system with length L. So τ_t is the *thermal transit time.* Conversely

$$\frac{1}{\tau_t} = \omega_t$$

is the number of times that the thermal signal traverses the system in a time unit. Hence, it is the *thermal transit frequency.* With this notation the first criterion (2.33) can be rewritten in the following two equivalent forms

$$\tau_{t,e} \gg 1/\omega_{p,e}$$

$$\omega_{p,e} \gg \omega_{t,e} \qquad (2.34)$$

These conditions require that the thermal electron transit time is much longer than the period of the plasma electron oscillation (note that when we are talking about "much longer or much larger than" we do not worry about a factor 2π) so that the electrons perform a huge number ($\tau_{t,e}\omega_{p,e} \gg 1$) of oscillations around their equilibrium position during the time it takes a thermal electron signal to cross the system. The plasma electron frequency $\omega_{p,e}$ must be much larger than the thermal electron transit frequency $\omega_{t,e}$. Plasma behaviour is observed only on time scales longer than the plasma period and on length scales larger than the Debye length.

2.5 Weakly coupled plasmas

The screening of the electric charges that we have discussed in the previous Section has physical meaning only if a large number of particles is present in the screening cloud. If there is only a handful of screening particles, the charge q_i remains unscreened in most directions at any given time. It is therefore useful to compute the number of particles (electrons) N_D that lie within the Debye sphere of q_i, and take part in the screening. The number N_D of particles in the Debye sphere is

$$N_D = \frac{4\pi}{3}(\lambda_D)^3 n_e = 1.7 \times 10^7 (k_B T/1\mathrm{keV})^{3/2} (n_e/10^{19}\mathrm{m}^{-3})^{-1/2}$$

For a plasma at $k_BT = 1\text{keV}$ with $n_e = 10^{19}\text{m}^{-3}$, N_D takes the value 1.7×10^7. (The clever reader has noted some ambiguity in the definition of N_D, he/she knows very well that the particle density (both electron and ion density) is a function of the radial distance, so that we should use a volume integral. We do not want to do that and therefore we interpret n_e as an average density!)

We can now quantify the condition for effective screening to occur: it requires N_D to be large: $N_D \gg 1$. Dropping from N_D the factor $\frac{4\pi}{3}$ we obtain the plasma parameter Λ^*

$$\Lambda^* = n_e \lambda_D^3 \qquad (2.35)$$

The second condition for an ionized gas to be called a plasma is then

$$\Lambda^* \gg 1 \qquad (2.36)$$

A large number of particles in the Debye sphere means that the average interparticle distance is much smaller than the Debye length. In a plasma with electron number density n_e, the distance between a particle and its nearest neighbour is roughly $n_e^{-1/3}$. Hence

$$n_e^{-1/3} \ll \lambda_D^*$$

Sometimes the plasma parameter is defined as $1/\Lambda^*$ and denoted by g:

$$g = 1/\Lambda^* = \frac{1}{n_e \lambda_D^3} \qquad (2.37)$$

The second criterion (2.36) for the definition of a plasma is then

$$g \ll 1 \qquad (2.38)$$

The second condition for an ionized gas to be called a plasma can also be obtained by imposing that the potential energy of a typical particle due to its nearest neighbour is much smaller than its kinetic energy so that *the plasma is weakly coupled.* As noted in the previous paragraph, the distance between a particle and its nearest neighbour is roughly $n_e^{-1/3}$. The average potential energy Φ of a particle due to its nearest neighbour is, in absolute value,

$$\mid \Phi \mid \sim \frac{e^2}{\epsilon_0 r} \sim \frac{n_e^{1/3} e^2}{\epsilon_0}$$

Recall that the average kinetic energy of a particle due to random motions is $3k_BT/2 = 3m_e v_t^2/2$. The condition that the potential energy of a typical particle due to its nearest neighbour is much smaller than its kinetic energy is

$$\frac{n_e^{1/3} e^2}{\epsilon_0} \ll k_BT$$

It is straightforward to show that this condition is equivalent to (2.36).

If $\lambda_D \simeq n_e^{-1/3}$ we have single independent particles. It is instructive to rewrite the plasma parameter Λ^* as

$$\Lambda^* = \frac{(\epsilon_0 k_B)^{3/2}}{e^3}\frac{T^{3/2}}{n^{1/2}} \qquad (2.39)$$

Thus weakly coupled plasmas are relatively hot and not too dense. The classic expression of $\Lambda^* = n_e\lambda_D^3$ seems to suggest otherwise. The observation that Λ^* measures the number of electrons in the Debye sphere (a sphere of radius λ_D) does not imply that Λ^* can be increased by squeezing more particles in a fixed container, that is by increasing the density. Because $\lambda_D \sim n^{-1/2}$, a decrease in density results in more particles in the Debye sphere.

2.6 Damping of plasma oscillations

Let us go back to the plasma oscillations. In Section 2 we have looked at the dynamic response of a plasma toward deviations from charge neutrality. In our attempt to understand the dynamics we used a very simple mathematical model for describing the plasma. This idealized mathematical model made our life very easy. However we had to pay a price for that easy life. Remember that we found harmonic plasma oscillations of the electron fluid about its equilibrium position that go on for ever. Everybody knows that in real life there is dissipation and transfer of momentum due to friction. The same is true in the case of the plasma electron oscillations. So we have to add more physics to our very simple mathematical model of Section 2 of this Chapter. We do not want to overdo it; we do not want to get things too complicated already in the second Chapter. What we want right now is friction. Due to the collisions of the electrons with the ions and neutral particles there is a loss of momentum of the electrons. Hence the collisions act as a friction force or drag force on the electrons. The classic assumption is that the loss of momentum of the electrons due to the collisions is proportional to the relative velocity of the electron fluid with respect to the background ions and neutrals, which are assumed to be relatively stationary. The drag force is then

$$\vec{F}_{fr} = -\overline{\nu}_{e,x} n_e m_e \vec{v}_e \qquad (2.40)$$

where $\overline{\nu}_{e,x}$ is an average composite collision frequency for momentum transfer from electrons to ions and neutrals. An expression for $\overline{\nu}_{e,i}$ will be derived in Section 7 of the present Chapter. There are two options. The first one is to go back to the set of equations (2.21) and add the friction force (2.40) to the right hand side of the equation of motion and then repeat the calculation carried out there. The second option is to note that the set of equations (2.21) led us to the electron plasma oscillations. If we denote the displacement of a electron fluid element as $\vec{\xi}$ then the equation of motion for an electron fluid element oscillating about its equilibrium position, is

$$\frac{d^2\vec{\xi}}{dt^2} + \omega_{p,e}^2\vec{\xi} = 0$$

Hence we can start from this equation and add the drag force (2.40). The equation of motion then takes the form

$$\frac{d^2\vec{\xi}}{dt^2} + \overline{\nu}_{e,x}\frac{d\vec{\xi}}{dt} + \omega_{p,e}^2\vec{\xi} = 0$$

The solution to this linear second order differential equation is

$$\xi(\vec{t}) = \xi_0 \exp(-\frac{\overline{\nu}_{e,x}}{2}t)\exp\left\{i(\omega_{p,e}^2 - \overline{\nu}_{e,x}^2/4)^{1/2}t\right\}$$

which represents a damped oscillation for

$$\frac{\overline{\nu}_{e,x}}{2} < \omega_{p,e}$$

We do not want to admit it but basically we are very conservative and want to keep what we have already got. We found plasma oscillations in Section 2, and we want to keep plasma oscillations; we can live with slight damping but refuse that friction would convert oscillations into purely exponentially decaying solutions. If the plasma oscillations are to be only slightly damped, it is necessary that the electron neutral collision frequency be much smaller than the electron plasma frequency,

$$\begin{aligned} \overline{\nu}_{e,x} &\ll \omega_{p,e} \\ \omega_{p,e}\tau_{e,x} &\gg 1 \end{aligned} \qquad (2.41)$$

Otherwise, the electrons will not be able to behave in an independent way, but will be forced to be in complete equilibrium with the ions and the neutrals, and the medium can be treated as a neutral gas. The preceding inequality constitutes therefore *the third criterion for the definition of a plasma.*

$$\tau_{e,x} = 1/\overline{\nu}_{e,x}$$

represents the average time an electron travels between collisions with ions and neutrals. It implies that the average time between collisions of electrons with ions and neutrals must be large compared to the period of the plasma oscillations. In this analysis the collisions of electrons with other electrons have been left out. These collisions are associated with the thermal motions of the electrons and electron pressure. The effect of electron pressure on electron plasma oscillation will be studied in Section 2 of the following Chapter.

2.7 Collisions

In the previous Section we have seen that the collisions of the electrons with the ions and neutrals cause a drag force on the electron fluid, which results in slightly damped plasma electron oscillations. The drag force is proportional to the relative momentum of the electron fluid to the ions and neutrals. The factor of proportionality is the average collisions frequency. This Section focusses on collisions of electrons with ions and derives an expression for the average collision frequency of electrons with ions, $\overline{\nu}_{e,i}$.

The words collision and interaction will be used as synonyms. The notion of a collision as a physical contact between bodies looses its utility in the microscopic world. On a microscopic scale a collision between particles must be regarded as an interaction between the force fields associated with each of the interacting particles.

Collisions of charged particles in a plasma are of two types: collisions with other charged particles and collisions with neutral atoms and molecules. Electrically charged particles interact with one another according to Coulomb's law. This Coulomb interaction is a *long-range interaction* so that one particle interacts simultaneously with a large number of particles. The force fields associated with neutral particles are significantly strong only within the electronic shells of the particles and obviously *short-range*. A neutral particle only occasionally interacts with another particle. The collisions of charged particles with other charged particles dominate by far over collisions with neutral particles in high-temperature plasmas where the degree of ionization is high. As a matter of fact Coulomb collisions dominate over collisions with neutrals in any plasma that is even just a few percent ionized. Only if the ionization level is very low ($< 10^{-3}$) can neutral collisions dominate. Moreover a plasma becomes almost fully ionized at electron temperatures above about 1 eV ($T_e = 11605K$). Thus, the case of collisions with neutrals is only of concern to a person interested in *low-temperature plasmas*.

Collisions can be divided in *elastic* and *inelastic* collisions. In elastic collisions the colliding particles retain their identity and remain in the same internal energy state. There is neither creation nor annihilation of particles and there is conservation of mass, momentum and energy. In inelastic collisions the internal energy state of some or all of the colliding particles may be changed. The colliding particles can loose their identity and particles may be created, as well as destroyed. Inelastic collisions lead to recombination, ionization and excitation.

Collisions are important because they tend to make the distribution functions isotropic and cause collisional transport. There is a particular interest in collisions between electrons and ions in a plasma because these collisions impede the acceleration of electrons in response to an electric field applied along (or in the absence of) a magnetic field. Without such collisions, the electrons would be accelerated indefinitely by an applied electric field, so that a infinitesimally small voltage would be sufficient to drive a large current through a plasma, at least in the direction along the magnetic field. In practice, the acceleration of electrons is impeded by collisions with non-accelerated particles, in particular the ions, which, because of their much larger mass, are relatively unresponsive to the applied electric field. Collisions between electrons and ions, acting in this way to limit the current that can be driven by an electric field, give rise to an important plasma quantity, namely *the plasma electrical resistivity*, denoted by $\tilde{\eta}$.

Binary collisions are described in terms of the impact parameter b and the angle of deflection χ. The relation between these two quantities depends on the type of interaction, i.e. the nature of the inter-particle force. In particular we shall see that this relation for a Coulomb interaction is quite different from that for a hard billiard ball interaction.

The angular distribution of the scattered particles is conveniently described in terms of cross-sections. Cross-sections are usually defined in terms of a beam of particles incident on a centre of force (target particle). Most widely used are the differential cross-section, total cross-section and cross-section for momentum.

Let γ denote the number of particles that hit the target per unit time, which obviously is also the number of particles scattered by the target per unit time. The *differential scattering cross-section* $\frac{d\sigma}{d\Omega}$ is defined as

$$\frac{d\sigma}{d\Omega} = \frac{\frac{d\gamma}{d\Omega}}{\frac{d\gamma}{d\sigma}} \tag{2.42}$$

and is the ratio of the number of scattered particles per unit solid angle and per unit time over the number of incident particles per unit area and per unit time. Note that the number of incident particles per unit area and per unit time is the incident particle flux.

The *total scattering cross-section* σ_t is the number of particles scattered per unit time and per unit incident flux, in all directions from the scattering centre. It is obtained by integrating $\frac{d\sigma}{d\Omega}$ over the entire solid angle:

$$\sigma_t = \int \frac{d\sigma}{d\Omega} d\Omega \qquad (2.43)$$

The *momentum transfer cross-section* is defined as

$$\sigma_m = \int \text{fractional loss of momentum of particles} \times \frac{d\sigma}{d\Omega} d\Omega \qquad (2.44)$$

The differential cross-section $\frac{d\sigma}{d\Omega}$ is related to the impact parameter b and the angle of deflection χ as

$$\frac{d\sigma}{d\Omega} = \frac{b}{\sin\chi} \left| \frac{db}{d\chi} \right| \qquad (2.45)$$

The absolute value of $db/d\chi$ is used here, because χ normally decreases when b increases and $\frac{d\sigma}{d\Omega}$ is a positive quantity.

Let us now look at the simple case of an elastic collision of a light particle (e.g. an electron) with a heavy electrically neutral particle. The light particle essentially bounces off the heavy particle, the two particles retain their identity and remain in the same internal energy state. The light particle may loose any fraction of its initial momentum, depending on the angle at which it rebounds. We start with this simple case because it will help us to explain how Coulomb collisions differ from billiard ball collisions. The heavy particle or target is regarded as a fixed and hard (that is, perfectly elastic) sphere of radius R. The target is hit by a uniform parallel beam of light particles, which we treat as points. Since the beam is uniform, we have a constant incident particle flux

$$f = \frac{d\gamma}{d\sigma}, \quad \gamma = f\sigma_t$$

where σ_t is the cross-sectional area presented by the target, namely $\sigma_t = \pi R^2$. Consider now one of the incoming particles that impinges on the target with velocity w and impact parameter b. It hits the target at an angle α to the normal given by $b = R\sin\alpha$. The particle is deflected through an angle χ so that $\chi + 2\alpha = \pi$. The relation between the impact parameter b and the angle of deflection χ is

$$b = R\cos\frac{\chi}{2}$$

Small angle deflections with $\chi \to 0$ correspond to an impact parameter $b \to R$, while large angle deflections with $\chi \to \pi$ correspond to an impact parameter $b \to 0$. Note that $\chi = \pi$ is a head on collision with impact parameter $b = 0$. The important observation to make is that $0 \leq b \leq R$, and there is no scattering for $b > R$. The calculation of the differential cross-section is straightforward. We use (2.45) and $b = R\cos\frac{\chi}{2}$ to find

$$\frac{d\sigma}{d\Omega} = \frac{R^2}{4}$$

Note that the differential cross-section for collisions with a fixed hard sphere is a constant which means that the scattering is isotropic.

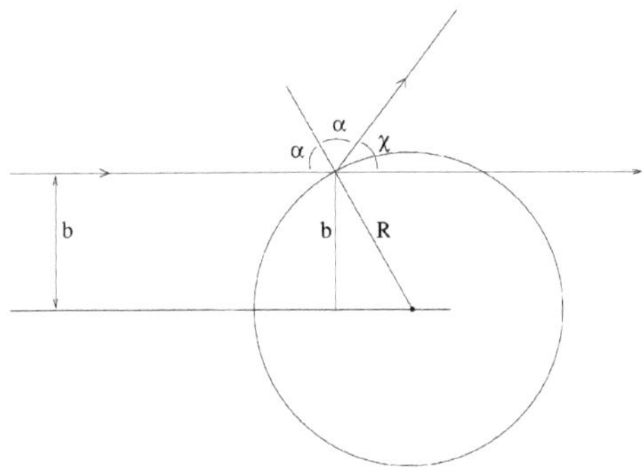

Figure 2.5: *Collision of a light point-like particle with a heavy particle.*

Let us now look at collisions of electrons with ions. We shall show that a naive substitution of the Coulomb potential in the calculation leads to a false prediction. It will also point to the necessity of taking into account the organized or collective behaviour of many charged particles brought about by the long-range interactions. The reader is advised to have a look at his notes of classical mechanics on scattering in a central force field and to familiarize himself again with the famous Rutherford formula for the differential cross-section.

We are looking at a light electron with mass m_e and charge $q_e = -e$ and a heavy ion with mass m_i and charge $q_i = Ze$. The heavy ion remains at rest. The initial velocity of the electron is $\vec{w}$. In the strictly mathematical sense $\vec{w}$ is the velocity at infinity where the potential energy of the electron is zero. The total energy E of the electron is

$$E = \frac{1}{2} m_e w^2$$

In absence of the ion the trajectory of the electron is a straight line. The central force field of the ion is

$$\vec{F}(r) = \frac{k}{r^2}\vec{1}_r, \quad k = \frac{q_e q_i}{4\pi\epsilon_0} = \frac{-Ze^2}{4\pi\epsilon_0}$$

and causes the trajectory of the electron to deviate from a straight line. The electron's trajectory is still planar but it is now a hyperbola (since $E > 0$!). The original straight line is an asymptote and the ion (the scattering centre) a focal point of this hyperbola. The distance from the position of the ion to the original straight line is the *impact parameter* b as is indicated on Fig. 2.6. The equation of the hyperbola is

$$r(e \cos\theta + 1) = l$$

in polar coordinates, and

$$\frac{(x - ae)^2}{a^2} - \frac{y^2}{b^2} = 1$$

in Cartesian coordinates. The origin of the system of coordinates is the position of the heavy ion; the x-axis is the straight line through the position of the ion and the intersection of the

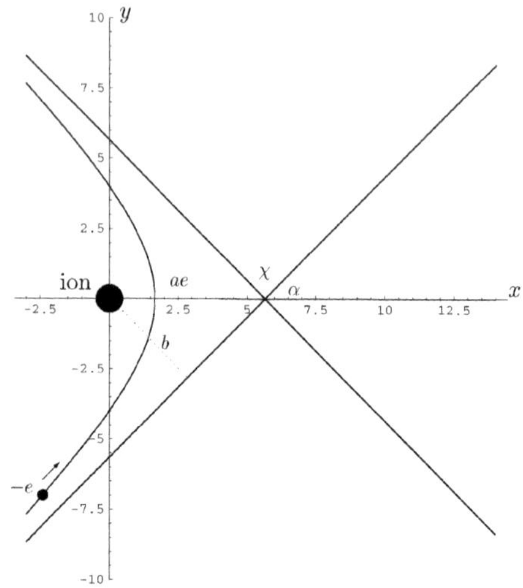

Figure 2.6: *The orbit of an electron undergoing a Coulomb collision with a fixed ion of charge Ze.*

asymptotes. Since the Coulomb force is attractive, the trajectory of the particle is the left half of the hyperbola. The constants in the equations are

$$\begin{aligned} l &= \frac{J^2}{m \mid k \mid} \\ J &= mbw \\ e^2 &= \frac{2lE}{m \mid k \mid} + 1 > 1 \\ a &= \frac{l}{e^2 - 1} = \frac{\mid k \mid}{2E} = \frac{\mid k \mid}{m_e w^2} = \frac{Ze^2}{4\pi\epsilon_0 m_e w^2} \\ b^2 &= a^2(e^2 - 1) \end{aligned}$$

A translation of the origin along the $x-$axis with $\tilde{x} = x - ae$ leads to the canonical form of the equation of the hyperbola:

$$\frac{\tilde{x}^2}{a^2} - \frac{y^2}{b^2} = 1$$

The equation of the asymptotes is

$$y = \pm\frac{b}{a}\tilde{x} = \pm\frac{b}{a}(x - ae)$$

Since we are looking at the left part of the hyperbola we have $x \leq ae$ and $y \leq 0$ before collision and $y \geq 0$ after collision. We denote the angle that the original straight line; i.e. the asymptote, makes with the $x-$axis as α so that

$$\tan\alpha = \frac{b}{a}, \quad \tan^2\alpha = \frac{b^2}{a^2} = e^2 - 1$$

The particle is deflected through an angle χ so that

$$\chi + 2\alpha = \pi, \quad e^2 - 1 = \cot^2\frac{\chi}{2}$$

The relation between the impact parameter b and the angle of deflection χ is

$$b = a\cot\frac{\chi}{2}, \quad \tan\frac{\chi}{2} = \frac{a}{b} \tag{2.46}$$

Since a can be expressed in terms of $E = m_e w^2/2$ and $\mid k \mid$, the impact parameter b is determined by the energy at infinity, the strength of the interaction and the scattering angle. The quantity a has an obvious physical meaning. Let us denote the impact parameter for scattering at a right angle $\chi = \pi/2$ as b_0. Since $\tan\pi/4 = 1$ it follows that

$$b_0 = a = \frac{Ze^2}{4\pi\epsilon_0 m_e w^2} \tag{2.47}$$

In what follows we shall write the relation (2.46) between the impact parameter b and the angle of deflection χ with a replaced with b_0.

Large angle deflections are deflections with a scattering angle $\chi \geq \pi/2$. For a head-on collision $\chi = \pi$ and $b = 0$. Large angle deflections have an impact parameter $b \leq b_0$. Let us now look at small angle deflections. We have that

$$\chi \to 0 \;\Rightarrow\; \tan\frac{\chi}{2} \to 0 \;\Rightarrow b \to \infty$$

and

$$\chi_{\rm min} = 0 \quad\Leftrightarrow\quad b_{\rm max} = +\infty$$

This is a truly amazing result as it implies that the electron we are looking at is simultaneously having small angle Coulomb interactions with all of the other particles in the plasma. This is definitely collective behaviour, obviously it is too much of that. For Coulomb interactions in a plasma $b \in [0, \infty[$ while for collisions with a hard sphere of radius R, $b \in [0, R]$.

We now use (2.45) and (2.46) to find the differential cross-section for Coulomb collisions to be

$$\boxed{\frac{d\sigma}{d\Omega} = \frac{b_0^2}{4\sin^4(\chi/2)} = \frac{b_0^2}{(1-\cos\chi)^2}} \tag{2.48}$$

This is the famous Rutherford cross-section, originally derived by Rutherford for the scattering of α particles by atomic nuclei.

The *total scattering cross-section* σ_t (2.43) is obtained by integrating the differential cross-section over the entire solid angle:

$$\sigma_t = \int \frac{d\sigma}{d\Omega} d\Omega = 2\pi b_0^2 \int_{\chi_{\rm min}}^{\pi} \frac{\sin\chi}{(1-\cos\chi)^2} d\chi$$

The lower integration limit has been written as $\chi_{\min}$ on purpose. With the use of the substitution $y = 1 - \cos\chi$, $y_{\min} = 1 - \cos\chi_{\min}$ we can evaluate the integral over χ as

$$\int_{\chi_{\min}}^{\pi} \frac{\sin\chi}{(1-\cos\chi)^2} d\chi = \frac{1}{y_{\min}} - \frac{1}{2} = \frac{1}{2}\cot^2\frac{\chi_{\min}}{2}$$

The total cross-section for Coulomb collisions is

$$\sigma_t = \pi b_0^2 \cot^2 \frac{\chi_{\min}}{2} \qquad (2.49)$$

For $\chi_{\min} = 0$ the total cross-section σ_t diverges to infinity. This is due to the fact that the impact parameter b becomes unbounded for small angle defections. The particles with very small deflection angles contribute to make σ_t infinite! In order to find the total cross-section for large angle deflections we have to put $\chi_{\min} = \pi/2$ and $\cot(\chi_{\min}/2) = 1$ so that

$$\sigma_{t_{\chi \geq \pi/2}} = \pi b_0^2$$

Let us now look at the cross-section for momentum transfer. The momentum of the electron before interaction is $m_e w$. After interaction the electron is scattered at an angle χ and the momentum of the particle in the direction of incidence is $m_e w \cos\chi$. Therefore, the relative loss of momentum of the electron in the direction of incidence is $1 - \cos\chi$. The cross-section for momentum transfer (2.44) is obtained by integrating $(1-\cos\chi)\frac{d\sigma}{d\Omega}$ over the entire solid angle:

$$\sigma_m = \int (1-\cos\chi)\frac{d\sigma}{d\Omega} d\Omega = 2\pi b_0^2 \int_{\chi_{\min}}^{\pi} \frac{\sin\chi}{(1-\cos\chi)} d\chi$$

With the use of the substitution $y = 1 - \cos\chi$, $y_{\min} = 1 - \cos\chi_{\min}$ we can evaluate the integral over χ as

$$\int_{\chi_{\min}}^{\pi} \frac{\sin\chi}{(1-\cos\chi)} d\chi = \int_{y_{\min}}^{2} y^{-1} dy = \ln\left(\frac{2}{y_{\min}}\right) = 2\ln\left(\frac{1}{\sin(\chi_{\min}/2)}\right)$$

The cross-section for transfer of momentum for Coulomb collisions is

$$\sigma_m = 4\pi b_0^2 \ln \frac{1}{\sin(\chi_{\min}/2)} \qquad (2.50)$$

For $\chi_{\min} = 0$ the cross-section for momentum transfer σ_m diverges logarithmically to infinity. The origin of the logarithmic divergence are the scattering events with large impact parameters and small scattering angles. We can find the cross-section for transfer of momentum for large angle deflections by putting $\chi_{\min} = \pi/2$ and $\sin\frac{\chi_{\min}}{2} = \sqrt{2}/2$ so that

$$\sigma_{m_{\chi \geq \pi/2}} = 4\pi b_0^2 \ln\sqrt{2}$$

The use of the Coulomb potential (2.29) leads to infinite values for both σ_t and σ_m. This is too bad. The cross-section for transfer of momentum is virtually proportional to the electric resistivity of the plasma; experiments tell us that ordinary plasmas are characterized by finite (as a rule very small) values of resistivity, not infinite ones. The infinite results are due to the absence of a cut-off value for the impact parameter b. Small values of the angle of deflection χ correspond to large values of b, and for $\chi_{\min} = 0$ we must have $b_{\max} = +\infty$. In order to obtain finite and meaningful values for σ_t and σ_m, it is necessary to modify the treatment of the interactions of charged particles and introduce a cut-off value for the impact parameter.

The disease is now well understood; a simple picture of binary Coulomb scattering cannot correctly describe the behaviour of charged particles interacting at large distances in the plasma. The cure for the disease is also known, instead of the bare Coulomb potential (2.29) we need to use the Debye screened potential (2.31). A charged particle inside a plasma is shielded by a cloud of particles of opposite sign and this Debye shielding results in an exponential decrease in the electric potential for $r > \lambda_D$. Thus, the maximum value of the impact parameter should be taken to be λ_D so that

$$b_{\max} = \lambda_D \quad \Rightarrow \quad \cot(\chi_{\min}/2) = \frac{\lambda_D}{b_0}, \quad \sin(\chi_{\min}/2) = (1 + (\lambda_D/b_0)^2)^{-1/2}$$

With this cut-off value for $b_{\max}$ and $\chi_{\min}$ the total cross-section σ_t (2.49) is

$$\sigma_t = \pi \lambda_D^2 \qquad (2.51)$$

The total cross-section for large angle deflections was found earlier $\sigma_{t_{\chi \geq \pi/2}} = \pi b_0^2$. The total cross-section for small angle deflections ($\chi \leq \pi/2$) is simply

$$\sigma_{t_{\chi \leq \pi/2}} = \sigma_t - \sigma_{t_{\chi \geq \pi/2}} = \pi(\lambda_D^2 - b_0^2)$$

and the ratio of the total cross-section for small angle deflections to the total cross-section for large angle deflections is

$$\frac{\sigma_{t_{\chi \leq \pi/2}}}{\sigma_{t_{\chi \geq \pi/2}}} = \frac{\lambda_D^2}{b_0^2} - 1 = \Lambda^2 - 1$$

The quantity Λ is defined as

$$\Lambda = \frac{\lambda_D}{b_0}$$

With the use of the definition of λ_D (2.30) and that of b_0 (2.47) we find

$$\Lambda = \frac{\lambda_D}{b_0} = \frac{12\pi}{Z} n \lambda_D^3 = \frac{9N_D}{Z} >> 1 \qquad (2.52)$$

where N_D is the number of particles in the Debye sphere. Λ is a very large number so that the cross-section for small angle deflections is far larger than that for large angle deflections.

Since $\sin(\chi_{\min}/2) = (1 + (\lambda_D/b_0)^2)^{-1/2}$ the cross-section for transfer of momentum σ_m (2.50), is

$$\sigma_m = 4\pi b_0^2 \ln \left\{ 1 + \frac{\lambda_D^2}{b_0^2} \right\}^{1/2} = 4\pi b_0^2 \ln \Lambda = \frac{Z^2 e^4}{4\pi \epsilon_0^2 m_e^2 w^4} \ln \Lambda \qquad (2.53)$$

$\ln \Lambda$ is called the Coulomb logarithm. The cross-section for transfer of momentum due to large angle deflections was found earlier $\sigma_{m_{\chi \geq \pi/2}} = 4\pi b_0^2 \ln \sqrt{2}$. The cross-section for transfer of momentum for small angle deflections $(\chi \leq \pi/2)$ is simply

$$\sigma_{m_{\chi \leq \pi/2}} = \sigma_m - \sigma_{m_{\chi \geq \pi/2}} = 4\pi b_0^2 (\ln \Lambda - \ln \sqrt{2})$$

and the ratio of the cross-section for transfer of momentum for small angle deflections to that for large angle deflections is

$$\frac{\sigma_{m_{\chi \leq \pi/2}}}{\sigma_{m_{\chi \geq \pi/2}}} = \frac{\ln \Lambda - \ln \sqrt{2}}{\ln \sqrt{2}} = \frac{\ln \Lambda}{\ln \sqrt{2}} - 1 >> 1$$

The function $\ln \Lambda$ varies slowly over a large range of variation of parameters on which Λ depends. Values of $\ln \Lambda$ in natural and laboratory plasmas are 12 in a gas discharge, 14 in the Earth's magnetosphere, 18 in a fusion reactor, 21 in the solar corona, 26 in the solar wind and in the Van Allen belts.

Although less pronounced than in the case of the total cross-section σ_t it is again the cross-section for transfer of momentum for small angle deflections that is far larger than that for large angle deflections. The large number of small angle deflections (weak interactions) is much more important than the small number of large angle deflections (strong interactions)! This is really collective behaviour at work!

In what follows we shall denote the cross-section for transfer of momentum for collisions of electrons with ions as σ_{ei}. The collision frequency ν_{ei} for (light) electrons striking (heavy) ions can be obtained from the usual relation between collision frequency ν_{ei} and collision cross-section σ_{ei}:

$$\nu_{ei} = n_i \sigma_{ei} w$$

With the use of (2.53)

$$\boxed{\nu_{ei} = \frac{n_i Z^2 e^4}{4\pi\epsilon_0^2 m_e^2 w^3} \ln \Lambda \qquad (2.54)}$$

The collision frequency varies with electron velocity as w^{-3}, i.e. the more fast-moving the electron the less frequently it collides with ions! In order to define an *average electron collision frequency*, it is necessary to evaluate the frictional force or drag force on a distribution of electrons moving through essentially stationary ions, namely

$$\boxed{\vec{F}_{fr} = -m_e \int \nu_{ei} \vec{w} f_e(\vec{w}) d^3\vec{w} = -n_e m_e \overline{\nu}_{e,i} < \vec{w} > \qquad (2.55)}$$

where the average is over the distribution of electron velocities. For present purposes, we suppose that the drifting electrons have a shifted Maxwell-Boltzmann distribution (2.13) with non-zero mean velocity $< \vec{w} >= \vec{v}$ with respect to the heavy immobile ions. We take this non-zero mean velocity to be in the $z-$direction, i.e. $\vec{v} = v\vec{1}_z$:

$$f_e(\vec{w}) = \frac{n_e}{(\sqrt{2\pi} v_{t,e}^3)} \exp\left(-\frac{|\vec{w} - \vec{v}|^2}{2v_{t,e}^2}\right)$$

Furthermore we assume that the motion of the electrons as a whole with respect to the heavy ions is slow in comparison with random motions of the electrons so that $|\vec{v}| \ll v_{t,e}$. We use

$$\begin{aligned}
\frac{|\vec{w}-\vec{v}|^2}{2v_{t,e}^2} &= \frac{|\vec{w}|^2}{2v_{t,e}^2} - \frac{\vec{w}.\vec{v}}{v_{t,e}^2} + \frac{|\vec{v}|^2}{2v_{t,e}^2} \\
\exp(-\frac{|\vec{w}-\vec{v}|^2}{2v_{t,e}^2}) &\approx \exp(-\frac{w^2}{2v_{t,e}^2})\exp(\frac{\vec{w}.\vec{v}}{v_{t,e}^2}) \\
&\approx (1+\frac{\vec{w}.\vec{v}}{v_{t,e}^2})\exp(-\frac{w^2}{2v_{t,e}^2})
\end{aligned}$$

to rewrite the distribution function $f_e(\vec{w})$ as

$$f_e(\vec{w}) = f_{e,0}(\vec{w})(1+\frac{\vec{w}.\vec{v}}{v_{t,e}^2})$$

where $f_{e.0}(\vec{w})$ is the standard unshifted Maxwell-Boltzmann distribution (2.10).

With our choice of the coordinate system with the $z-$axis along the average velocity of the electron stream the frictional force $\vec{F}_{fr}$ (2.55) has only a $z-$component:

$$\begin{aligned}
F_{fr,z} &= -m_e \underbrace{\int \nu_{ei} w_z f_{e,0}(\vec{w}) d^3\vec{w}}_{I_1} - m_e \frac{v_z}{v_{t,e}^2} \underbrace{\int \nu_{ei} w_z^2 f_{e,0}(\vec{w}) d^3\vec{w}}_{I_2} \\
&= -m_e I_1 - m_e \frac{v_z}{v_{t,e}^2} I_2
\end{aligned}$$

where I_1 and I_2 are obvious abbreviations for the integrals over $\vec{w}$ which we evaluate with the use of the expression for ν_{ei}.

$$I_1 = \frac{n_e n_i Z^2 e^4 \ln\Lambda}{(\sqrt{2\pi}v_{t,e})^3 4\pi\epsilon_0^2 m_e^2} \underbrace{\int \frac{w_z}{w^3} \exp\left(-\frac{w^2}{2v_{t,e}^2}\right) d^3\vec{w}}_{A}$$

We use spherical coordinates $d^3\vec{w} = w^2 \sin\theta \; dw \; d\theta \; d\phi, \;\; w_z = w\cos\theta$ to find that $A = 0$ so that $I_1 = 0$.

$$I_2 = \frac{n_e n_i Z^2 e^4 \ln\Lambda}{(\sqrt{2\pi}v_{t,e})^3 4\pi\epsilon_0^2 m_e^2} \underbrace{\int \frac{w_z^2}{w^3} \exp\left(-\frac{w^2}{2v_{t,e}^2}\right) d^3\vec{w}}_{B}$$

Use again spherical coordinates and also $y = w^2/2v_{t,e}^2$ to calculate B and find $B = \frac{4\pi}{3}v_{t,e}^2$. The expression (2.55) for $F_{fr,z}$ takes the form

$$F_{fr,z} = -n_e m_e v_z \frac{n_i Z^2 e^4 \ln\Lambda}{6\pi\sqrt{2\pi}v_{t,e}^3 \epsilon_0^2 m_e^2}$$

Comparing this expression for $F_{fr,z}$ with $F_{fr,z} = -n_e m_e \overline{\nu}_{e,i} < w_z >= -n_e m_e \overline{\nu}_{e,i} v_z$ we find that

$$\overline{\nu}_{e,i} = \frac{n_i Z^2 e^4 \ln\Lambda}{6\pi\sqrt{2\pi}v_{t,e}^3 \epsilon_0^2 m_e^2}$$

Now use $v_{t,e}^2 = k_B T_e/m_e$ to make the dependence of $\overline{\nu}_{e,i}$ on the electron temperature explicit:

$$\overline{\nu}_{e,i} = \frac{\sqrt{2} n_i Z^2 e^4 \ln \Lambda}{12\pi^{3/2} \epsilon_0^2 k_B^{3/2} \sqrt{m_e}} \frac{1}{T_e^{3/2}} \tag{2.56}$$

Hence, $\overline{\nu}_{e,i}$ varies inversely with $T_e^{3/2}$ and is independent of the ion mass. In Chapter IV we shall see that the electrical resistivity $\tilde{\eta}$ of a plasma is related to $\overline{\nu}_{e,i}$ as

$$\tilde{\eta} = \frac{m_e \overline{\nu}_{e,i}}{n_e e^2} \tag{2.57}$$

so that

$$\tilde{\eta} = \frac{\sqrt{2} n_i \sqrt{m_e} Z^2 e^2 \ln \Lambda}{12\pi^{3/2} \epsilon_0^2 k_B^{3/2} n_e} \frac{1}{T_e^{3/2}} \tag{2.58}$$

This expression shows that the resistivity of a fully ionized plasma is independent of its density and varies inversely with $T_e^{3/2}$. As the temperature of a plasma is raised, its resistivity drops rapidly. Plasmas at (very) high temperature are most likely to be *perfectly conducting* or *collisionless*, meaning that their electrical resistivity is negligible. However, the decrease in resistivity with increasing temperature has a severe disadvantage for *Ohmic heating* of plasmas. Ohmic heating is a simple method for heating plasmas which involves passing a current through the plasma to dissipate some energy in heat. The rate by which a plasma is heated by this method is ηj^2 per unit volume. For fixed j, the heating rate drops as the temperature rises, so much so that Ohmic heating is usually considered impractical at fusion temperatures.

2.8 Larmor frequency and Larmor radius

So far I have not said anything about magnetic fields although I have tried to convince the reader that magnetic fields can play an important role in plasmas. It is high time I do something about that. Many of the plasmas in nature and in the laboratory occur in the presence of magnetic fields. An important property of magnetic fields is that they confine plasmas. Every plasma has the natural tendency to disperse. Unless there is a restraining force, the energetic particles that compose the plasma will travel away from their initial positions at high velocity and the plasma ceases to exist. Often, plasmas have low densities and collisions are rather infrequent so that a particle travelling outwards is not likely to be deflected by a collision. Nevertheless, both in laboratory and in space, diffuse plasmas can be sustained. They are prevented from dispersing by magnetic fields which act on the charged particles through the Lorentz force. For this reason it is necessary to have basic insight in the dynamics of charged particles in magnetic fields. Here we consider the simplest possible situation, which is the motion of single charged particles in a magnetic field. As a matter of fact this is not really plasma physics but rather single particle dynamics in an external magnetic field. We consider the case of a charged particle with mass m and charge q moving in a spatially uniform magnetic field which does not vary in time. The equation of motion is

$$m \frac{d\vec{w}}{dt} = q\vec{w} \times \vec{B} \tag{2.59}$$

For what follows, it is helpful to decompose $\vec{w}$ as

$$\vec{w} = \vec{w}_{\parallel} + \vec{w}_{\perp}$$

where

$$\vec{w}_{\parallel} = (\vec{w}.\vec{1}_B)\vec{1}_B, \quad \vec{w}_{\perp} = \vec{w} - \vec{w}_{\parallel}$$

are the velocity respectively along the field lines and in the plane normal to the magnetic field lines. $\vec{1}_B = \vec{B}/B$ is the unit vector along the magnetic field line. With this decomposition, it is easy to show that

$$w^2, \quad w_{\parallel}, \quad w_{\perp} = \text{constants of motion} \tag{2.60}$$

The magnitude of the velocity perpendicular to the field: $w_{\perp}$ is the most useful constant for describing the motion in a plane perpendicular to the magnetic field. We now take the z-axis parallel to $\vec{B}$ so that

$$\vec{B} = (0, 0, B_z)^t, \quad \vec{w}_{\parallel} = (0, 0, w_z)^t, \quad \vec{w}_{\perp} = (w_x, w_y, 0)^t, \quad \vec{w} \times \vec{B} = (w_y B_z, -w_x B_z, 0)^t$$

The components of the equation of motion (2.59) are

$$\frac{dw_x}{dt} = \pm\omega_c w_y, \quad \frac{dw_y}{dt} = \mp\omega_c w_x, \quad \frac{dw_z}{dt} = 0$$

where

$$\boxed{\omega_c = \frac{|\, q \,|}{m} B \qquad (2.61)}$$

has the dimension of a frequency. The upper (lower) signs apply to the ions (electrons). The x- and y-components (i.e. the components perpendicular to the magnetic field) of the equation of motion can be combined to

$$\frac{d^2 w_x}{dt^2} = -\omega_c^2 w_x, \quad \frac{d^2 w_y}{dt^2} = -\omega_c^2 w_y$$

These are the standard equations for an harmonic oscillator of frequency ω_c. The solution for w_x is

$$w_x = w_{\perp} \sin(\omega_c t + \phi)$$

where $w_{\perp}$ is the constant speed of the particle in the (x, y) plane and ϕ is a constant of integration that fixes the velocity components the (x, y)-plane at $t = 0$ according to

$$\tan\phi = \frac{w_x(0)}{w_y(0)}$$

To determine w_y we substitute the solution for w_x in the left-hand side of $dw_x/dt = \pm\omega_c w_y$ and find for the ions

$$w_y = w_{\perp} \cos(\omega_c t + \phi).$$

The general solution for the ions is

$$w_x = w_{\perp} \sin(\omega_c t + \phi), \quad w_y = w_{\perp} \cos(\omega_c t + \phi), \quad w_z = w_{\parallel}$$

where of course $w_x^2 + w_y^2 = w_\perp^2$. Integration with respect to time t gives the coordinates of the ion:

$$x(t) = -\frac{w_\perp}{\omega_c}\cos(\omega_c t + \phi) + X_0, \quad y(t) = \frac{w_\perp}{\omega_c}\sin(\omega_c t + \phi) + Y_0, \quad z(t) = w_\parallel t + z_0 \tag{2.62}$$

X_0 and Y_0 are related to the initial position of the ion in the (x, y) plane according to

$$X_0 = x(0) + \frac{w_\perp}{\omega_c}\cos\phi, \quad Y_0 = y(0) - \frac{w_\perp}{\omega_c}\sin\phi$$

Similar expressions for w_x, w_y, x, y, z can be found for the electrons. Straightforward elimination of t from the equations for x and y gives

$$(x - X_0)^2 + (y - Y_0)^2 = r_L^2 \tag{2.63}$$

where

$$\boxed{r_L = \frac{w_\perp}{\omega_c}} \tag{2.64}$$

is the Larmor radius. In the (x, y) plane the solution simply represents a uniform motion with velocity $w_\perp$ in a circle with its centre at (X_0, Y_0) and radius r_L. The angular frequency $\omega_c =$ of the circular motion and the radius r_L of the circle in the x, y plane are called the Larmor frequency or the cyclotron gyration frequency and the Larmor radius respectively. The constants X_0 and Y_0 are the coordinates of the centre of the circular motion. Superimposed on this motion along the Larmor circle in a plane perpendicular to B is a uniform motion parallel to B so that the complete motion is along a helical line (see Fig. 2.7). The nature of this motion is easily understood if it is noted that the magnetic force is always perpendicular to the particle velocity, so that it does no work on the particle and the magnitude of the velocity is constant. The force is then of constant magnitude and at right angles to the velocity and the magnetic field, just what is needed to produce circular motion in the plane perpendicular to the field.

The angular frequency and the radius of the circular motion are

$$\boxed{\begin{aligned} \omega_{c,e} &= \frac{eB}{m_e}, \quad r_{L,e} = \frac{w_\perp}{\omega_{c,e}} = \frac{m_e w_\perp}{eB} \quad \text{for electrons} \\ \omega_{c,i} &= \frac{ZeB}{m_i}, \quad r_{L,ei} = \frac{w_\perp}{\omega_{c,i}} = \frac{m_i w_\perp}{eB} \quad \text{for ions} \end{aligned}} \tag{2.65}$$

These are the four fundamental quantities that characterize the dynamics of the electrons and the ions in a magnetic field. Note that

$$\frac{\omega_{c,e}}{\omega_{c,i}} = \frac{m_i}{m_e} >> 1 \tag{2.66}$$

so that the light electrons always rotate much faster than the heavy ions. In (2.66) the ions are protons. The direction of the gyration of the charged particle around the field direction can be deduced from the expression for v_x and v_y, or more easily by noting that the magnetic force is to the centre of the orbit. The conclusion is that electrons (negatively charged particles) rotate

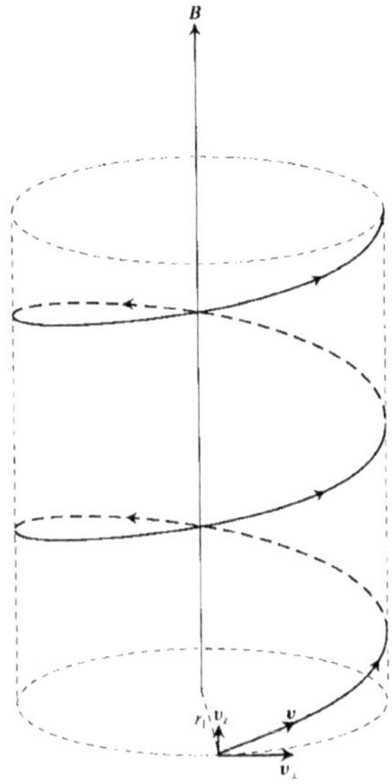

Figure 2.7: *Helical path of an electron in a uniform magnetic field.*

in a right hand sense when viewed along the field while the ions (positively charged particles) rotate in the opposite sense as is shown on Fig. 2.8 (see also Problem 12). Of course the reader is eager to find out whether he is dealing with a microscopic or macroscopic phenomenon. In order to find that out we have to replace the universal constants e, m_e and m_p by their numerical values $e = 1.6022 \times 10^{-19}$C, $m_e = 9.1094 \times 10^{-31}$kg, $m_p = 1.6726 \times 10^{-27}$kg. The results for the Larmor frequency of an electron and a proton are

$$\omega_{c,e} = 1.76 \times 10^{11} \frac{B}{1\text{Tesla}} \text{rad} \times \text{s}^{-1}, \quad \omega_{c,p} = 9.58 \times 10^{8} \frac{B}{1\text{Tesla}} \text{rad} \times \text{s}^{-1}$$

This shows us that the Larmor frequency is a microscopic frequency which is of the same order of magnitude as the plasma electron frequency for typical fusion plasmas.

For the radius of the Larmor circle of an electron we find

$$r_{L,e} = \frac{w_\perp}{\omega_{c,e}} = 7.6 \times 10^{-5} \frac{w_\perp}{1.3 \times 10^{7} m \times s^{-1}} (\frac{B}{1Tesla})^{-1} m$$

The normalizing velocity $1.3 \times 10^{7} ms^{-1}$ is that of an electron of 1 keV. The dependence of r_L on $w_\perp$ and B follows from the fact that the magnetic field provides the centripetal force which holds an otherwise free electron in a circular perpendicular motion. Thus r_L is the distance at which there is equilibrium:

$$\frac{m(w_\perp)^2}{r_L} = e w_\perp B$$

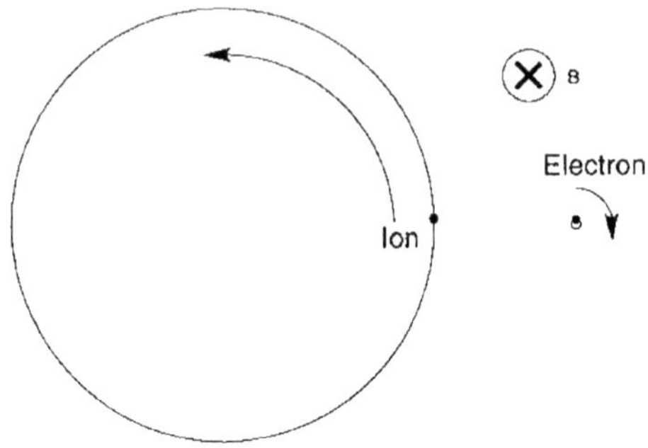

Figure 2.8: *Ion and electron gyro-motion in a magnetic field. For fixed energy, the ion's gyro-orbit is much larger than the electron's. '×' indicates that the magnetic field faces into the page.*

In a weak magnetic field, the force causing the electron to deviate from a straight line is small, and the radius of curvature r_L is correspondingly large. In a strong magnetic field the electron is forced to follow the magnetic field line on a tight helical trajectory. The dependence of r_L on $w_\perp$ follows from two considerations: the more energetic the electron, the more difficult it is for the magnetic field to deflect the electron, however, the Lorentz force itself is proportional to the electron perpendicular velocity. The quantities $\omega_{p,e}, \lambda_D, \omega_{c,e}, r_{L,e}$ characterize the high-frequency behaviour of a plasma dictated by the dynamics of the electrons.The quantities $\omega_{p,i}, \omega_{c,i}, r_{L,i}$ characterize the low-frequency behaviour of a plasma dictated by the dynamics of the ions.

The charged particles are tied to the magnetic field lines; they move in Larmor circles around the magnetic field lines with their Larmor frequency. Recall that collisions tend to make the distribution functions isotropic. Here we see that the magnetic field plays the role of collisions for the perpendicular motion, that is, perpendicular to the field, particles are confined to the vicinity of a field line executing a nearly isotropic motion if their gyroradius is much smaller than the plasma dimension. On the other hand, the motion of the charged particles along the magnetic field lines is unaffected by the magnetic field. The magnetic field introduces a preferred direction for the charged particles into the plasma and a strong magnetic field can cause anisotropy in the plasma with different dynamics perpendicular and parallel to the field lines!

Let us have a closer look at the notion of the electrons and ions being tied to the magnetic field lines. Individual particles have different random velocities. In order to understand the confining effect of the magnetic field on the particles we compute the Larmor radius (2.64) with $w_\perp$ replaced with the thermal velocity. The Larmor radius (2.64) for particles moving at their thermal speed is

$$r_{L,t} = \frac{v_t}{\omega_c} \qquad (2.67)$$

The ratio of the Larmor radii for electrons and ions is then

$$\frac{r_{L,t,e}}{r_{L,t,i}} = \frac{v_{t,e}}{v_{t,i}}\frac{\omega_{c,i}}{\omega_{c,e}} = \frac{v_{t,e}}{v_{t,i}}\frac{m_e}{m_i}$$

In case the electrons and the ions have the same temperature, i.e. $T_e \approx T_i$, then $v_{t,e}/v_{t,i} = (m_i/m_e)^{1/2}$ so that

$$\frac{r_{L,t,e}}{r_{L,t,i}} = \left(\frac{m_e}{m_i}\right)^{1/2} << 1 \tag{2.68}$$

The light electrons are always much closer tied to the magnetic field lines than the heavy ions. Let us now compare the Larmor gyration radius for both electrons and ions with the length scale L of the system. This is done by the use of the ratio δ

$$\delta = \frac{r_{L,t}}{L} \tag{2.69}$$

The particles are *magnetized* if

$$\boxed{\delta = \frac{r_{L,t}}{L} << 1} \tag{2.70}$$

In most cases electrons can be treated as magnetized; i.e. as effectively glued to the magnetic field lines. However, this is not obvious for the heavy ions. Since, because of (2.68) and (2.69)

$$\delta_i = \left(\frac{m_i}{m_e}\right)^{1/2} \delta_e >> \delta_e$$

magnetization of the ions requires a much stronger magnetic field than that needed for the magnetization of electrons. In Chapter IV we shall discuss ideal MHD. The most basic property of ideal MHD is the freezing of the magnetic field lines in the plasma. This property corresponds to the mathematical idealization

$$\boxed{\delta_e = 0, \quad r_{L,e} = 0, \quad \delta_i = 0, \quad r_{L,i} = 0} \tag{2.71}$$

This is OK for the very light electrons, but might cause problems for the heavy ions. Hence we can expect deviations from ideal MHD because of the ion dynamics. This possibility is taken into consideration in the model of Hall MHD.

The magnetization parameter δ (2.69) is defined as the ratio of two lengths. It can also be written as the ratio of two frequencies. Recall that $\tau_t = L/v_t$ and $\omega_t = 1/\tau_t$ are the thermal transit time and the thermal transit frequency respectively. We rewrite the expression for δ (2.69) as

$$\delta = \frac{r_{L,t}}{L} = \frac{\omega_t}{\omega_c} \tag{2.72}$$

The species are magnetized if

$$\boxed{\begin{aligned} \frac{\omega_t}{\omega_c} &<< 1 \\ \tau_t \omega_c &\gg 1 \end{aligned}} \tag{2.73}$$

These are two equivalent conditions which require the electrons and ions to perform a large number of rotations around the magnetic field lines during the time it takes a thermal signal to cross the system.

2.9 Recapitulation

"What have I done?
What am I doing?
What shall I do?"
Homer J. Simpson
The Simpsons

- Distribution functions for particles.
  The real plasma consisting of discrete particles is replaced with a smeared-out density distribution function in phase space. Instead of looking at individual particles it is convenient to adopt a kinetic description with distribution functions in phase space for the different species.

  The evolution in the phase space and in time is governed by the Boltzmann equation (2.9). The Boltzmann equation (2.9) supplemented with an equation for the collision term and the Maxwell equations (2.7) form a closed set of equations. Once $f_\alpha(\vec{r}, \vec{w}, t)$ is known, any macroscopic quantity, such as density, pressure, temperature, can be computed by integration of expressions involving $f_\alpha(\vec{r}, \vec{w}, t)$ over velocity space. In particular the kinetic energy of the random motions is used to define the temperature.

  The classic Maxwell-Boltzmann distribution function (2.10) is the time independent distribution function for an isolated uniform plasma in thermodynamic equilibrium in absence of external forces. It is time independent, uniform, isotropic and has zero average velocity. Generalizations to non-zero average velocity, anisotropy and local thermodynamic equilibrium are straightforward.

- Plasma oscillations and plasma frequency.
  Plasma oscillations are high-frequency motions of electrons due to deviations from charge neutrality in a cold plasma. The heavy ions do not participate in these high-frequency oscillations.

- Debye shielding length λ_D and first criterion for a plasma.
  The potential around a point charge in a warm plasma differs from the Coulomb potential because of the screening by the other charged particles. For distances smaller than λ_D the potential is essentially the Coulomb potential, for distances larger than λ_D the potential decreases exponentially.

  The first criterion for a plasma concerns overall charge neutrality. The volume occupied by the plasma must be much larger than a Debye sphere so that there is overall charge neutrality in the plasma. Hence the spatial length scale L of the plasma must be much longer than the Debye shielding length. This inequality implies that charge neutrality is a good approximation only over time spans much longer than the period of the plasma oscillation.

- Plasma parameter and second criterion for a plasma.
The second criterion for a plasma is that there are many particles in the Debye sphere so that there is effective screening and collective behaviour. This also means that the average inter-particle distance is much smaller than λ_D and that the plasma is weakly coupled in the sense that the average potential energy of a particle to its nearest neighbour is much smaller than its kinetic energy.

- Damping of plasma oscillations and third criterion for a plasma.
Collisions of electrons with heavy neutral particles cause a loss of momentum of the electrons. The plasma oscillations are damped. The electrons are not to be forced to be in complete equilibrium with the heavy particles but should be able to behave in an independent way. The plasma oscillations are to be only slightly damped by the collisions of the electrons with the heavy neutral particles. The frequency for collisions of electrons with neutrals must be much smaller than the electron plasma frequency.

- Collisions.
Collisions of charged particles with neutrals are negligible in plasmas, only collisions with other particles play a role. Individual collisions are described in terms of the impact parameter b and the angle of deflection χ. The relation between these two quantities is determined by the type of interaction. For Coulomb interactions the maximal value of b corresponding to the minimal value of $\chi = 0$ is ∞.

  The interaction of a beam of incident particles with a target is described in terms of cross-sections: the differential cross-section, the total cross-section and the cross-section for transfer of momentum.

  For Coulomb interactions both the total cross-section and the cross-section for transfer of momentum are infinite. This is due to the absence of a cut-off value for the impact parameter. The Debye shielding distance is a reasonable choice for this cut-off value. The large number of small angle collisions is far more important than the few large angle collisions.

  The average collision frequency and the plasma resistivity are inversely proportional to $T_e^{3/2}$.

- Larmor gyration.
The charged particles are tied to the magnetic field lines; they move in Larmor circles around the magnetic field lines with their Larmor frequency. Their motion along the magnetic field lines is unaffected by the magnetic field. In most cases electrons can be treated as magnetized; i.e. as effectively glued to the magnetic field lines. However, this is not obvious for the heavy ions. Hence we can anticipate deviations from ideal MHD because of the ion dynamics.

- High-frequency electron and low-frequency ion dynamics
The quantities $\omega_{p,e}, \lambda_D, \omega_{c,e}, r_{L,e}$ characterize the high-frequency behaviour of a plasma dictated by the dynamics of the electrons. The quantities $\omega_{p,i}, \omega_{c,i}, r_{L,i}$ characterize the low-frequency behaviour of a plasma dictated by the dynamics of the ions.

2.10 Problems

"You have to treat these professors carefully, Persse" said Angelica, with a sly smile. "You have to flatter them a bit."
Angelica Pabst to Persse McGarrigle
Small World
David Lodge

1. The elements of the matrix M are easily calculated when it is noted that

$$x_i' = x_i + w_i \Delta t, \quad w_i' = w_i + a_i \Delta t$$

It is then easy to see that the matrix M can be written with four 3×3 blocks as

$$M = \begin{bmatrix} M_1 & M_2 \\ & \\ M_3 & M_4 \end{bmatrix}$$

Show that all elements on the diagonal of M are equal to 1 and that all off-diagonal elements either vanish or are proportional to Δt. (B1986).

2. The Maxwell-Boltzmann distribution function (2.10) is uniform and isotropic and as a consequence it makes sense to define a distribution function of (the absolute value of) speeds $F(w)$ with $w = \mid \vec{w} \mid$. $F(w)$ is defined so that the number of particles per unit volume having velocity between w and $w + \Delta w$ is $F(w)\Delta w$. Show that

$$F(w) = w^2 \int_0^{2\pi} d\phi \int_0^{\pi} d\theta \sin\theta f(\vec{w})$$

and derive its explicit expression for a Maxwell-Boltzmann distribution (2.10). (B1986).

3.
- Compute the average speed (or average value of $w = \mid \vec{w} \mid$) v_{av} for a Maxwell-Boltzmann distribution (2.10) with

$$v_{av} = < w > = < \mid \vec{w} \mid > = \frac{1}{n} \int_0^{\infty} wF(w)dw$$

Compute for Maxwell-Boltzmann distribution (2.10) the most probable speed v_{mp} as the speed for which $F(w)$ attains its maximum.
- Compare these two speeds with the root-mean-square speed v_{rms} defined as

$$v_{rms} = (< w^2 >)^{1/2} = (3k_B T/m)^{1/2}$$

and note that

$$v_{mp} < v_{av} < v_{rms}$$

(B1986)

4. Derive the equation for the perturbation of the number density of the electrons $n_{e,1}$ in a cold non-magnetic plasma when the ions are immobile:

$$\left\{\frac{\partial^2}{\partial t^2}+\frac{e^2 n_{e,0}}{m_e \epsilon_0}\right\} n_{e,1}=0$$

5. Derive the equation for the perturbation of the number density of the electrons $n_{e,1}$ in a cold non-magnetic plasma when the ions are allowed to move:

$$\left\{\frac{\partial^2}{\partial t^2}+\frac{e^2 n_{e,0}}{m_e \epsilon_0}\frac{m_i+m_e}{m_i}\right\} n_{e,1}=0$$

6. Find the plasma frequencies $\omega_{p,e}$, $\omega_{p,i}$, the Debye lengths λ_{De}, λ_{Di}, the plasma parameters Λ_e^*, Λ_i^*, and the thermal velocities $v_t=(kT/m)^{1/2}$ for electrons and ions (protons) under the following conditions

- a tokamak with $n_e=n_i=10^{19}m^{-3}$, $T_e=T_i=10^8K$
- the earth's magnetosphere with $n_e=n_i=10^{10}m^{-3}$, $T_e=T_i=10^3K$
- the centre of the Sun with $n_e=n_i=10^{32}m^{-3}$, $T_e=T_i=15\times10^6K$
- the solar corona with $n_e=n_i=10^{14}m^{-3}$, $T_e=T_i=10^6K$
- the solar wind with $n_e=n_i=10^7m^{-3}$, $T_e=T_i=10^5K$
- the atmosphere of a neutron star with $n_e=n_i=10^{22}m^{-3}$, $T_e=T_i=10^7K$

(D1990, S1994)

7. Consider Debye shielding in a 1-dimensional Cartesian system so that $\nabla^2=d^2/dx^2$. Find the solution for the Debye shielded potential for a plasma with $T_e=T_i$ and $Z=1$ and for a plasma with different electron and ion temperatures $T_e\neq T_i$ and general Z. In the latter case show that the Debye length is

$$\lambda_D=\left(\frac{\epsilon_0 k_B T}{n_0 e^2(1+ZT_e/T_i)}\right)^{1/2}$$

(GR2000)

8. Find the expression (2.32) for the total charge!

9. Show that the condition that a plasma is weakly coupled is indeed equivalent to the second condition for an ionized gas to be called a plasma.

10. Derive the relation between the differential cross-section and the impact parameter and the angle of deflection:

$$\frac{d\sigma}{d\Omega}=\frac{b}{\sin\chi}\mid\frac{db}{d\chi}\mid$$

(B1986)

11. Derive the equation for the hyperbolic orbit of an electron with mass m_e, charge $-e$ and velocity $\vec{w}$ at infinity under the influence of the Coulomb attraction of an immobile massive ion of mass m_i and charge Ze. Start from the equation of conservation of energy for a conservative force and the equation of conservation of angular momentum. First use the position of the massive ion as the origin of the coordinate system to obtain the equation for the hyperbole in polar coordinates. Subsequently use a translation of the origin to obtain the canonical form of the equation in Cartesian coordinates. (KB)

12. Derive the relation between the impact parameter and the angle of deflection for Coulomb collisions!

13. Compute the cross-section for transfer of momentum σ_m for collisions of a beam of pointlike particles with a hard sphere of radius R. (B1986)

14. Let $F(\chi)$ be a function of the angle of deflection χ. Since $\frac{d\sigma}{d\Omega}$ is an angular distribution function, it can be used as a weighting function to calculate the mean value of any function $F(\chi)$ of the angle χ as

$$< F(\chi) >= \frac{\int F(\chi)\frac{d\sigma}{d\Omega}d\Omega}{\int \frac{d\sigma}{d\Omega}d\Omega}$$

Use this definition to relate σ_m and σ_t as

$$\sigma_m = \sigma_t < 1 - \cos\chi >$$

and interpret σ_m/σ_t as the average value of momentum loss per particle.
Compute σ_m/σ_t for

- the collisions of point like particles with a hard sphere
- Coulomb interactions
- Debye shielded Coulomb interactions.

(B1986)

15. Use spherical coordinates $d^3\vec{w} = w^2 \sin\theta \; dw \; d\theta \; d\phi, \quad w_z = w\cos\theta$ to show that $A = 0$ and hence $I_1 = 0$ in Section 2.7.

16. Use again spherical coordinates and also $y = w^2/2v_t^2$ to calculate B in Section 2.7 and find $B = \frac{4\pi}{3}v_t^2$.

17. The electron mean-free path for collisions with ions is defined as

$$\lambda_{mfp} = w\tau_{ei} = \frac{w}{\nu_{ei}}$$

where $\tau_{e,i} = 1/\nu_{e,i}$ is the time an electron travels between two collisions with an ion. Show that for Debye shielded Coulomb collisions

$$\frac{\lambda_{mfp}}{\lambda_D} \approx \frac{\Lambda}{\ln\Lambda} >> 1$$

Interpret this result! (GR2000)

18. Show that

$$w^2, \; w_{\|}, \; w_{\perp}$$

are constants of motion for a charged particle in a straight constant magnetic field.

19. Determine the motion of an electron in a constant magnetic field; i.e. find expressions for w_x, w_y, w_z, x, y, z as we have done for the ions in Section 2.8. What can you say about the direction of the gyration of the ions and electrons?

20. Calculate the electron and proton gyrofrequencies and the gyroradii of the particles moving at their thermal velocities under the conditions

    - a tokamak with $B = 10^4 G$
    - the earth's magnetosphere with $B = 10^{-2} G$
    - the solar corona with $B = 1G$
    - the solar wind with $B = 10^{-5} G$
    - the atmosphere of a neutron star with $B = 10^{12} G$

    Compare the gyrofrequencies with the plasma frequencies and the gyroradii with the Debye lengths! (S1994)

21. Show that

$$\frac{\omega_{p,e}^2}{\omega_{c,e}} = \frac{\omega_{p,i}^2}{\omega_{c,i}}$$

22. Determine the motion of a charged particle in a constant electrostatic and magnetostatic field and show that it consists of three parts:

    - a constant acceleration $q\vec{E}_{\|}/m$ along the $\vec{B}$ field
    - a rotation about the direction of $\vec{B}$ at the cyclotron frequency
    - an electromagnetic drift velocity $\vec{w}_E = (\vec{E} \times \vec{B})/B^2$

    (B1986)

Dr. Irving Langmuir (1881-1957) is an American chemist and physicist who won the Nobel Prize in chemistry in 1932. Dr. Langmuir coined the term plasma as applied to ionized gases, studied the electrostatic plasma oscillations which bear his name (Langmuir waves), developed the theory of Langmuir probes, and applied his research to develop numerous inventions for General Electric, particularly in the field of lighting.

Ludwig Boltzmann (1844 - 1906).
Boltzmann was awarded a doctorate from the University of Vienna in 1866 for a thesis on the kinetic theory of gases supervised by Josef Stefan.
His personality certainly had a major impact on the direction that his career took and personal relationships, where he was always very soft-hearted, played a big part. He suffered from an alternation of depressed moods with elevated, expansive or irritable moods. Indeed his physical appearance, being short and stout with curly hair, seemed to fit his personality. His fiancée called him her "sweet fat darling".
Boltzmann's fame is based on his invention of statistical mechanics. This he did independently of Willard Gibbs. Their theories connected the properties and behaviour of atoms and molecules with the large scale properties and behaviour of the substances of which they were the building blocks. Boltzmann obtained the Maxwell-Boltzmann distribution in 1871, namely the average energy of motion of a molecule is the same for each direction. He was one of the first to recognise the importance of Maxwell's electromagnetic theory.
In 1884 the work of Josef Stefan was developed by Boltzmann who showed how Josef Stefan's empirical T^4 law for black body radiation, formulated in 1879, could be derived from the principles of thermodynamics. Boltzmann worked on statistical mechanics using probability to describe how the properties of atoms determine the properties of matter. In particular his work relates to the Second Law of Thermodynamics which he derived from the principles of mechanics in the 1890s.
Boltzmann's ideas were not accepted by many scientists. In 1904 Boltzmann visited the World's Fair in St Louis, USA. He lectured on applied mathematics and then went on to visit Berkeley and Stanford. Unfortunately he failed to realise that the new discoveries concerning radiation that he learnt about on this visit were about to prove his theories correct.
Boltzmann continued to defend his belief in atomic structure and in a 1905 publication Populäre Schriften he tried to explain how the physical world could be described by differential equations which represented the macroscopic view without representing the underlying atomic structure. :- May I be excused for saying with banality that the forest hides the trees for those who think that they disengage themselves from atomistics by the consideration of differential equations.
Attacks on his work continued and he began to feel that his life's work was about to collapse despite his defence of his theories. Depressed and in bad health, Boltzmann committed suicide just before experiment verified his work. On holiday with his wife and daughter at the Bay of Duino near Trieste, he hanged himself while his wife and daughter were swimming. However the cause of his suicide may have been wrongly attributed to the lack of acceptance of his ideas. We will never know the real cause which may have been the result of mental illness causing his depression.

http://www-history.mcs.st-andrews.ac.uk/history/Mathematicians/
Article by: J. J. O'Connor and E. F. Robertson

Chapter 3

Fluid equations for mass, momentum and energy

"You are one of the rare people who can separate your observation from your preconception. You see what is, where most people see what they expect."
Lee to Samuel Hamilton
East of Eden
John Steinbeck

In Chapter I have made it clear that the emphasis of this course is on Magnetohydrodynamics (MHD for short). In the previous Chapter we have seen that plasmas are most accurately described by particle distribution functions in phase space. The spatial and temporal evolution of these distribution functions are governed by the Boltzmann equation (2.9). This is a partial differential equation in a 7-dimensional space. Chapman and Enskog were the first to obtain a solution to the Boltzmann equation in 1916 for a mono-atomic gas close to thermodynamic equilibrium. However, in plasma physics we are confronted with a far more difficult situation. The full set of Boltzmann-Maxwell equations (2.7) - (2.9) provides a very detailed and complete description of plasma behaviour. At one end of the spectrum, it contains microscopic information about the orbits of the individual charged particles on the very short cyclotron time scale and the gyro-radius length scale. At the other end, it accurately describes the macroscopic behaviour of large astrophysical and fusion plasmas. The complexity arising from this wide range of plasma physics information makes it virtually impossible to solve the Boltzmann-Maxwell equations (2.7) - (2.9) in any nontrivial situation. The collision term makes life even more complicated. Usually this collision term is written as an integral involving the distribution function itself, so that Boltzmann equation (2.9) is really an integro-differential equation. You do not have to be a big genius to understand that it is horrendously difficult to obtain a solution to the Vlasov equation for collisionless plasmas or even worse to the Boltzmann equation under realistic plasma physics conditions. This realization has led to the development of simpler mathematical models with a narrower range of applicability. Magnetohydrodynamics is such a model. Psychology has probably a name for it; if you cannot get it, you pretend you do not want it. Actually, it often makes sense to not want to solve the Boltzmann equation (2.9) since its solution would give too much and too detailed information. In many cases our interest lies in macroscopic quantities like density, temperature and pressure and how these macroscopic quantities vary in space and time. These macroscopic quantities are obtained as moments of the distribution function. Clearly, it will be simpler to investigate their evolution than that of the distribution function.

The macroscopic moments are quantities that we are already familiar with from fluid and gas dynamics. The resulting theory falls into the domain of fluid theory. The aim of the present Chapter is to derive part of the hydrodynamic theory of plasmas known as Magnetohydrodynamics. MHD is a macroscopic, non-relativistic theory that is concerned with global phenomena in magnetic plasmas. It gives an accurate description of many of the complicated interactions of magnetic fields with the plasmas of the sun and stars. MHD can be viewed as classical fluid dynamics with the additional complication that the fluid is electrically conducting. A short way of introducing the equations of MHD is to write down the constitutive equations of classical fluid dynamics and to add the terms and equations due to the magnetic field. The problem is that it is not obvious which terms are important and should be retained in the equations. At best, only handwaving arguments are used. In addition, it is also unclear under which conditions the equations are a valid idealization of the plasma physics world. So we have decided not to take this short way; we have the desire to understand what we are doing.

In order to understand more fully the physical content of MHD as well as the limitations of the model, the equations of MHD are derived starting from basic principles. The derivation follows a standard procedure which has found its way into the more recent textbooks on plasma physics and MHD. Fluid moments are calculated from a general kinetic model and a number of (simplifying) assumptions are made to obtain closure of the system of mathematical equations. The starting point for the derivation of the MHD equations is the full set of the Maxwell equations (2.7) coupled with a kinetic model of the plasma, described by a Boltzmann equation (2.9) for each species. However, MHD is a single-fluid theory of plasmas. It is already a further approximation to a more general hydrodynamic theory of plasmas, the *multi-fluid theory* of plasmas. In the following Section we are going to derive the multi-fluid equations. Then we specialize to a plasma consisting of electrons and ions and write down the two-fluid equations. In the final Section we derive the equations of MHD or single-fluid theory.

3.1 Multi-fluid theory

In this Section we derive the equations for the evolution of the basic macroscopic moments or quantities for particles of type α, i.e. number density $n_\alpha(\vec{x},t)$ (or mass density $\rho_\alpha(\vec{x},t)$), bulk velocity $\vec{v}_\alpha(\vec{x},t)$, the pressure tensor $P_{\alpha,ik}$. In the previous Section we have diagnosed the two main reasons that make the Boltzmann equation (2.9) so awfully difficult to solve: (i) too many independent variables, (ii) the collision term. The cure for the first disease is straightforward. The mathematical description is simplified by removing $\vec{w}$ by taking moments of the Boltzmann equation (2.9). This is exactly what is done in the derivation of the multi-fluid theory. When we take a moment of the Boltzmann equation (2.9) we actually transform a single equation for $f_\alpha(\vec{r},\vec{w},t)$ in seven variables $(\vec{r},\vec{w},t)$ into a fluid equation in four variables $(\vec{r},t)$. In principle we should take an infinite number of moments in order to obtain a closed set of equations. This is not possible from a practical point of view and we have to introduce assumptions in order to truncate the infinite set of moment equations and achieve closure of the finite set of equations. The truncated set of equations is by design focused on a specific regime of plasma physics. It gives a less detailed description of a plasma but we do not worry about this loss of microscopic information contained in $f_\alpha(\vec{r}, \vec{w}, t)$. We are interested in macroscopic quantities and these are found by integration over velocity space anyway. The cure for the second disease is to use as much as possible quantities that show collisional invariance so that the details of the collision term are not important, but global conservation laws for elastic collisions can be used.

The collision term in the right hand side of the Boltzmann equation is abbreviated as C_α and decomposed as

$$\left(\frac{\partial f_\alpha}{\partial t}\right)_{\text{coll}} = C_\alpha = \sum_\beta C_{\alpha,\beta}$$

where $C_{\alpha,\beta}$ is the contribution to C_α of collisions of particles of species α with particles of species β. In what follows we only consider elastic collisions (and neglect all effects of inelastic collisions such as ionization, recombination, and so on). For elastic collisions we have the following conservation laws:

- Conservation of particles of like and unlike particles:

$$\int C_{\alpha,\alpha} d^3\vec{w} = \int C_{\alpha,\beta} d^3\vec{w} = \int C_{\beta,\alpha} d^3\vec{w} = 0$$

- Conservation of momentum and energy between like particle collisions:

$$\int m_\alpha \vec{w} C_{\alpha,\alpha} d^3\vec{w} = 0 \ , \ \int \frac{1}{2} m_\alpha w^2 C_{\alpha,\alpha} d^3\vec{w} = 0$$

- Conservation of total momentum and energy between unlike particle collisions:

$$\int (m_\alpha C_{\alpha\beta} + m_\beta C_{\beta\alpha})\, \vec{w} d^3\vec{w} = 0 \ , \ \int \frac{1}{2} (m_\alpha C_{\alpha\beta} + m_\beta C_{\beta\alpha})\, w^2 d^3\vec{w} = 0$$

The equations of multi-fluid theory are obtained by taking the moments of the Boltzmann equation (2.9) that correspond to mass, momentum and energy for each species.

$$\int \{\text{Boltzmann's equation}\} \ d\vec{w} \ \Rightarrow \ \text{conservation of mass}$$

$$\int \{\text{Boltzmann's equation}\} \ \vec{w} \ d\vec{w} \ \Rightarrow \ \text{conservation of momentum}$$

$$\int \{\text{Boltzmann's equation}\} \ \frac{w^2}{2} \ d\vec{w} \ \Rightarrow \ \text{conservation of energy}$$

At this point the reader is advised to go back to Chapter II and refresh his memory with the definitions of $n_\alpha(\vec{r}, t), \vec{v}_\alpha(\vec{r}, t), Q_\alpha(\vec{r}, t), \vec{j}_\alpha(\vec{r}, t)$. Quantities for particles of species α are obtained by taking an average over velocity space. For a quantity $g(\vec{w}, \vec{r}, t)$ the average $< g(\vec{r}, t) >_\alpha$ is defined as

$$< g(\vec{r},t) >_\alpha = \frac{1}{n_\alpha(\vec{r},t)} \int g(\vec{w},\vec{r},t) f_\alpha(\vec{r},\vec{w},t) d^3\vec{w} \tag{3.1}$$

Conservation of particles

The zeroth order moment of the Boltzmann equation (2.9) for particles of type α is

$$\int \{\text{Boltzmann's equation}\}\, d^3\vec{w}$$

or

$$\underbrace{\int \frac{\partial f_\alpha}{\partial t} d^3\vec{w}}_{T_1} + \underbrace{\int \vec{w}.\nabla_x f_\alpha d^3\vec{w}}_{T_2} + \underbrace{\int \frac{\vec{F}_\alpha}{m_\alpha}.\nabla_w f_\alpha d^3\vec{w}}_{T_3} = \underbrace{\int C_\alpha d^3\vec{w}}_{=0}$$

Note that x_i, w_j, and t are independent coordinates. This means that the integration over $d^3\vec{w}$ and differentiation with respect to the time $\partial/\partial t$ and with respect to the spatial coordinates ∇_x can be interchanged. In Problem 1 you are asked to show that

$$T_1 = \frac{\partial n_\alpha}{\partial t}, \quad T_2 = \nabla_x.(n_\alpha \vec{v}_\alpha), \quad T_3 = 0$$

With the use of these results we obtain *the equation for conservation of particles of species α*

$$\frac{\partial n_\alpha}{\partial t} + \nabla_x.(n_\alpha \vec{v}_\alpha) = 0, \quad \frac{\partial \rho_\alpha}{\partial t} + \nabla_x.(\rho_\alpha \vec{v}_\alpha) = 0 \tag{3.2}$$

This is the continuity equation for particles of species α. It tells us that in the absence of any interaction processes which create or annihilate particles of this species, the particle number density and also the mass density and the charge density remain unchanged. The continuity equation is the first fluid equation of our multi-fluid plasma. It couples the plasma density n_α, ρ_α to the fluid velocity $\vec{v}_\alpha$. Hence we need another equation for the fluid velocity $\vec{v}_\alpha$. Since $\vec{v}_\alpha$ is the first moment of the distribution function, this equation will result from the first order moment of the Boltzmann equation (2.9) for particles of type α.

Conservation of momentum

The first order moment of the Boltzmann equation (2.9) for particles of type α is

$$m_\alpha \int \{\text{Boltzmann's equation}\}\, \vec{w}\, d^3\vec{w}$$

or

$$\underbrace{\int \frac{\partial f_\alpha}{\partial t} m_\alpha w_k d^3\vec{w}}_{T_1} + \underbrace{\int m_\alpha w_k \vec{w}.\nabla_x f_\alpha d^3\vec{w}}_{T_2} + \underbrace{\int w_k \vec{F}_\alpha.\nabla_w f_\alpha d^3\vec{w}}_{T_3} = \int C_\alpha m_\alpha w_k d^3\vec{w}$$

In Problem 2 you are asked to show that

$$T_1 = \frac{\partial}{\partial t}(\rho_\alpha v_{\alpha,i}), \quad T_2 = \sum_{l=1}^{3} \frac{\partial}{\partial x_l}(\rho_\alpha < w_k w_l >_\alpha)$$

$< w_k w_l >_\alpha$ denotes the average of $w_k w_l$ over the particles of type α as defined in (3.1). The random velocity $\vec{u}_\alpha$ of the particles of type α with respect to their average velocity $\vec{v}_\alpha$ is defined in (2.4). These random velocities are important for the thermodynamic quantities (e.g. pressure, temperature). Of course $< \vec{u}_\alpha >_\alpha = 0$. We now use the random velocities to rewrite $< w_k w_l >_\alpha$ (see Problem 3) as

$$< w_k w_l >_\alpha = v_{\alpha,k} v_{\alpha,l} + < u_{\alpha,k} u_{\alpha,l} >_\alpha$$

Insert this result in T_2 and find

$$T_2 = \sum_{l=1}^{3} \frac{\partial}{\partial x_l} (\rho_\alpha v_{\alpha,k} v_{\alpha,l}) + \sum_{l=1}^{3} \frac{\partial}{\partial x_l} (\rho_\alpha < u_{\alpha,k} u_{\alpha,l} >_\alpha)$$

Not unexpectedly the first order moment of the Boltzmann equation contains terms of second order in the random velocities $\rho_\alpha < u_{\alpha,k} u_{\alpha,l} >_\alpha$. In order to figure out what these terms are, it is helpful to recall the definitions of pressure and temperature in an ideal gas

$$\begin{aligned} p &= n k_B T \\ k_B T &= \frac{m}{3} < u^2 > \\ p &= \frac{nm}{3} < u^2 > = \frac{\rho}{3} < u^2 > \end{aligned}$$

For an isotropic ideal gas all elements of the tensor $< u_i u_j >$ are zero with the exception of those on the diagonal which are equal: $< u_x^2 > = < u_y^2 > = < u_z^2 > = < u^2 > /3$. Hence the tensor $\rho_\alpha < u_{\alpha,i} u_{\alpha,l} >_\alpha$ is actually a generalization of pressure for a plasma. This tensor is called the *pressure tensor* and denoted as $P_{\alpha,kl}$:

$$P_{\alpha,kl} = \rho_\alpha < u_{\alpha,k} u_{\alpha,l} >_\alpha = \int m_\alpha f_\alpha(\vec{r}, \vec{w}, t)(w_k - v_{\alpha,k})(w_l - v_{\alpha,l}) d^3\vec{w} \qquad (3.3)$$

Often the pressure tensor is split into an isotropic pressure p_α and its anisotropic part $\Pi_{\alpha,kl}$:

$$\begin{aligned} P_{\alpha,kl} &= p_\alpha \delta_{kl} + \Pi_{\alpha,kl} \\ p_\alpha &= \frac{\rho_\alpha}{3} <| \vec{u}_\alpha |^2>_\alpha, \quad \Pi_{\alpha,kl} = P_{\alpha,kl} - p_\alpha \delta_{kl} \qquad (3.4) \end{aligned}$$

p_α is the isotropic pressure of the particles of type α. $\Pi_{\alpha,kl}$ is the stress tensor due to the anisotropy of the distribution function f_α. It is identically zero for an isotropic distribution function (f.e. a Maxwell-Boltzmann distribution (2.10)). Note also that the diagonal elements of $\Pi_{\alpha,kl}$ can be different from zero. $P_{\alpha,kl}$ is symmetric and can always be diagonalized by the principal axes theorem.

The internal energy $\mathcal{U}_\alpha$ is a measure for the kinetic energy of the particles of type α due to random motions. It is defined as

$$\mathcal{U}_\alpha = \frac{1}{2} <| \vec{u}_\alpha |^2>_\alpha \qquad (3.5)$$

and related to pressure and temperature

$$p_\alpha = n_\alpha k_B T_\alpha = \frac{2}{3}\rho_\alpha \mathcal{U}_\alpha = \rho_\alpha v_{t,\alpha}^2$$

The concept of an isotropic pressure makes sense when the distribution function f_α does not deviate strongly from isotropy. If f_α is strongly anisotropic with different temperatures parallel $T_{\alpha,\parallel}$ and perpendicular $T_{\alpha,\perp}$ to the magnetic field lines, the concept of an isotropic pressure is not of much use. For a bi-Maxwellian distribution function (2.14) with the z-axis along the magnetic field the pressure tensor is (see Problem 4)

$$P_{\alpha,x,x} = P_{\alpha,y,y} = n_\alpha k_B T_{\alpha,\perp}, \quad P_{\alpha,z,z} = n_\alpha k_B T_{\alpha,\parallel}, \quad k \neq l : \; P_{\alpha,k,l} = 0$$

We use the pressure tensor to rewrite

$$\int m_\alpha w_k \vec{w}.\nabla_x f_\alpha d^3\vec{w} = \sum_{l=1}^{3} \frac{\partial}{\partial x_l}\left(\rho_\alpha v_{\alpha,k} v_{\alpha,l} + p_\alpha \delta_{kl} + \Pi_{\alpha,kl}\right)$$

In Problem 5 you are asked to compute T_3. The result is

$$T_3 = -\rho_\alpha g_k - Q_\alpha E_k - \left(\vec{j}_\alpha \times \vec{B}\right)_k$$

We denote the right hand side as $\mu_{\alpha,k}$ and compute it with the use of the random velocities (see Problem 6)

$$\vec{\mu}_\alpha = \int C_\alpha m_\alpha \vec{w} d^3\vec{w} = m_\alpha \int C_\alpha \vec{u}_\alpha d^3\vec{w} = m_\alpha \sum_\beta \int C_{\alpha,\beta} \vec{u}_\alpha d^3\vec{w} \tag{3.6}$$

$\vec{\mu}_\alpha$ is the average transfer of momentum from all particles to the particles of type α because of collisions. It is solely due to the random motions. $m_\alpha \int C_{\alpha,\beta} u_{\alpha,k} d^3\vec{w}$ is the average transfer of momentum from the particles of type β to the particles of type α. Combining the results for the various terms we obtain *the equation for conservation of momentum for particles of species* α:

$$\begin{aligned} \frac{\partial}{\partial t}\left(\rho_\alpha v_{\alpha,k}\right) &+ \sum_{l=1}^{3} \frac{\partial}{\partial x_l}\left(\rho_\alpha v_{\alpha,k} v_{\alpha,l} + p_\alpha \delta_{kl} + \Pi_{\alpha,kl}\right) \\ &- \rho_\alpha g_k - Q_\alpha E_k - \left(\vec{j}_\alpha \times \vec{B}\right)_k = \mu_{\alpha,k} \end{aligned} \tag{3.7}$$

This equation relates the fluid velocity to density, pressure and the electromagnetic force. The electromagnetic force now acts on plasma elements and not on individual particles. The equation for conservation of momentum for particles of species α is very similar to the Navier-Stokes equation of conventional hydrodynamics. It can be seen as the Navier-Stokes equation with the electromagnetic force added. The electromagnetic force couples the equation for conservation of momentum to the full set of Maxwell's equations (2.7) and makes it very distinct from the Navier-Stokes equation of conventional hydrodynamics where we have only pressure and viscous forces acting on the fluid elements. The electromagnetic force also couples all the charged plasma components together. The reason is that the electric and magnetic fields in the electromagnetic force act on all charged components and that at the same time all charged components contribute to the electric and magnetic fields. Hence, if you plan to solve the equation of motion for one plasma component, you cannot escape solving the equation of motion for all of the plasma components. All plasma components depend on each other and on the electric and magnetic fields.

Dyadic notation can be used to obtain more compact equations. The equation for conservation of momentum for particles of species α can be written as

$$\frac{\partial}{\partial t}(\rho_\alpha \vec{v}_\alpha) + \nabla.(\rho_\alpha \vec{v}_\alpha \vec{v}_\alpha + p_\alpha \mathbf{I} + \mathbf{\Pi}_\alpha) - \rho_\alpha \vec{g} - Q_\alpha \vec{E} - \vec{j}_\alpha \times \vec{B} = \vec{\mu}_\alpha \tag{3.8}$$

$\mathbf{I}$ is the unit tensor. Let us have a good look at the equation for conservation of momentum. Recall that this is the first order moment of the Boltzmann equation (2.9) and note that it contains second order quantities: pressure and the stress tensor. We do not have any equations for these second order quantities so far. You have already guessed a way out of this deadlock; take second order moments of the Boltzmann equation (2.9) (how many do you need actually?). You are right. But think harder and see more trouble on the horizon: the second order moments of the Boltzmann equation (2.9) will contain third order terms and in order to have equations for these third order terms you need to take fourth order moments of the Boltzmann equation (2.9) and so on. The point is that by now you should be convinced that in principle you need an infinite number of moments of the Boltzmann equation (2.9). We cannot handle an infinite number of equations, so we need to truncate the system. The penalty for doing that is a system of equations that is not closed. The remedy to that disease is additional assumptions. The situation is even worse since we do not have an equation for $\vec{\mu}$.

Conservation of energy

The second order moment of the Boltzmann equation (2.9) for particles of type α is

$$m_\alpha \int \{\text{Boltzmann's equation}\} \ \frac{w^2}{2} \ d^3\vec{w}$$

or

$$\underbrace{\int \frac{m_\alpha w^2}{2}\frac{\partial f_\alpha}{\partial t} d^3\vec{w}}_{T_1} + \underbrace{\int \frac{m_\alpha w^2}{2} \vec{w}.\nabla_x f_\alpha d^3\vec{w}}_{T_2} + \underbrace{\int \frac{w^2}{2} \vec{F}_\alpha.\nabla_w f_\alpha d^3\vec{w}}_{T_3} = \underbrace{\int \frac{m_\alpha w^2}{2} C_\alpha d^3\vec{w}}_{\nu_{\alpha,k}}$$

A straightforward calculation (see Problem 7) shows that

$$T_1 = \frac{\partial}{\partial t}(\frac{\rho_\alpha v_\alpha^2}{2} + \rho_\alpha \mathcal{U}_\alpha)$$

$$T_2 = \nabla.\{(\frac{\rho_\alpha v_\alpha^2}{2} + \rho_\alpha \mathcal{U}_\alpha)\vec{v}_\alpha + \frac{\rho_\alpha < u_\alpha^2 \vec{u}_\alpha >_\alpha}{2}\} + \sum_{k=1}^{3} \frac{\partial}{\partial x_k}(\sum_{l=1}^{3} P_{\alpha,kl} v_{\alpha,l})$$

So we had it right; we predicted a third order term to appear in the second order moment of the Boltzmann equation (2.9) and there it is: $\frac{1}{2}\rho_\alpha < u_\alpha^2 \vec{u}_\alpha >_\alpha$. It is the flux $\vec{\Phi}_\alpha$ of kinetic energy of random motions of the particles of type α by their random motions, or in other words the flux of heat of the particles of type α by the random motions:

$$\vec{\Phi}_\alpha = \frac{1}{2}\rho_\alpha < u_\alpha^2 \vec{u}_\alpha >_\alpha = \frac{1}{2} m_\alpha \int f_\alpha \mid (\vec{w} - \vec{v}_\alpha) \mid^2 (\vec{w} - \vec{v}_\alpha) d^3\vec{w} \tag{3.9}$$

With this notation T_2 can be written as

$$T_2 = \nabla . \left((\frac{\rho_\alpha v_\alpha^2}{2} + \rho_\alpha \mathcal{U}_\alpha)\vec{v}_\alpha + \vec{\Phi}_\alpha \right) + \sum_{k=1}^{3}\sum_{l=1}^{3} \frac{\partial}{\partial x_k} (v_{\alpha,l} P_{\alpha,kl})$$

In Problem 8 you are asked to compute T_3. The result is $T_3 = -\rho_\alpha \vec{v}_\alpha . \vec{g} - \vec{j}_\alpha . \vec{E}$. We compute ν_α with the use of the random velocities (see Problem 9)

$$\nu_\alpha = \int C_\alpha \frac{m_\alpha w^2}{2} d^3\vec{w} = H_\alpha + \vec{v}_\alpha . \vec{\mu}_\alpha$$

$$H_\alpha = \int C_\alpha \frac{m_\alpha u_\alpha^2}{2} d^3\vec{w} \tag{3.10}$$

is the gain (or loss) in heat of the particles of type α due to collisions with the other particles. $\vec{v}_\alpha . \vec{\mu}_\alpha$ can be interpreted as work done by the momentum transferred during the collisions on the particles of type α.

On combining the results for the various terms we obtain *the equation for conservation of energy for particles of species* α:

$$\begin{aligned} \frac{\partial}{\partial t}\left(\frac{\rho_\alpha v_\alpha^2}{2} + \rho_\alpha \mathcal{U}_\alpha \right) &+ \nabla . \left((\frac{\rho_\alpha v_\alpha^2}{2} + \rho_\alpha \mathcal{U}_\alpha)\vec{v}_\alpha + \vec{\Phi}_\alpha \right) + \sum_{k=1}^{3}\sum_{l=1}^{3} \frac{\partial}{\partial x_k} (P_{\alpha,kl} v_{\alpha,l}) \\ &- \rho_\alpha \vec{v}_\alpha . \vec{g} - \vec{j}_\alpha . \vec{E} = \nu_\alpha = H_\alpha + \vec{v}_\alpha . \vec{\mu}_\alpha \end{aligned} \tag{3.11}$$

The momentum equation and the equation for total energy may be simplified by using the continuity equation to remove the contributions $\frac{\partial \rho_\alpha}{\partial t}$. Also, since ∇_w will not be used any longer, we denote ∇_x as ∇. We obtain (see Problem 10)

$$\begin{aligned} \rho_\alpha \frac{d_\alpha v_{\alpha,k}}{dt} &+ \sum_{l=1}^{3} \frac{\partial}{\partial x_l} (p_\alpha \delta_{kl} + \Pi_{\alpha,kl}) \\ &- \rho_\alpha g_k - Q_\alpha E_k - \left(\vec{j}_\alpha \times \vec{B} \right)_k = \mu_{\alpha,k} \end{aligned}$$

$$\begin{aligned} \rho_\alpha \frac{d_\alpha}{dt}\left(\frac{1}{2} v_\alpha^2 + \mathcal{U}_\alpha \right) &+ \sum_{k=1}^{3} \frac{\partial}{\partial x_k} \left(\Phi_{\alpha,k} + \sum_{l=1}^{3} v_{\alpha,l} P_{\alpha,kl} \right) \\ &- \rho_\alpha \vec{v}_\alpha . \vec{g} - \vec{j}_\alpha . \vec{E} = \nu_\alpha = H_\alpha + \vec{v}_\alpha . \vec{\mu}_\alpha \end{aligned}$$

$$\frac{d_\alpha}{dt} = \frac{\partial}{\partial t} + \vec{v}_\alpha . \nabla$$

is the total time derivative with respect to the average flow of the particles of type α. The use of this total time derivative enables us to obtain somewhat shorter versions of the conservation laws. However, we have to remember that it is different for each type α of particles. In the Section on one-fluid theory we shall introduce and use the total time derivative with respect to the average velocity of the plasma as a whole.

Conservation of internal energy

We now take the scalar product of the equation for the conservation of momentum with the average velocity $\vec{v}_\alpha$ and find the equation for the change in kinetic energy as

$$\rho_\alpha \frac{d_\alpha}{dt}\left(\frac{v_\alpha^2}{2}\right) + \sum_{k=1}^{3}\sum_{l=1}^{3} \frac{\partial P_{\alpha,kl}}{\partial x_k} v_{\alpha,l} - \rho_\alpha \vec{v}_\alpha . \vec{g} - \vec{j}_\alpha . \vec{E} = \vec{v}_\alpha . \mu_\alpha$$

When we subtract this equation from the equation for the change in total energy we find the equation for the change in internal energy

$$\rho_\alpha \frac{d_\alpha \mathcal{U}_\alpha}{dt} + \sum_{l=1}^{3}\sum_{k=1}^{3} P_{\alpha,kl} \frac{\partial v_{\alpha,l}}{\partial x_k} + \nabla . \vec{\Phi}_\alpha = H_\alpha \qquad (3.12)$$

The equation the change in internal energy can be written in a more tractable form by using

$$\mathcal{U}_\alpha = \frac{3p_\alpha}{2\rho_\alpha}$$

$$P_{\alpha,kl} = p_\alpha \delta_{kl} + \Pi_{\alpha,kl}$$

$$\sum_{l=1}^{3}\sum_{k=1}^{3} P_{\alpha,kl} \frac{\partial v_{\alpha,l}}{\partial x_k} = -\frac{p_\alpha}{\rho_\alpha}\left(\frac{\partial}{\partial t} + \vec{v}_\alpha . \nabla\right)\rho_\alpha + \sum_{l=1}^{3}\sum_{k=1}^{3} \Pi_{\alpha,kl} \frac{\partial v_{\alpha,l}}{\partial x_k}$$

The resulting equation is

$$\frac{d_\alpha p_\alpha}{dt} - \frac{5}{3}\frac{p_\alpha}{\rho_\alpha}\frac{d_\alpha \rho_\alpha}{dt} = \frac{2}{3}\left\{-\nabla . \vec{\Phi}_\alpha - \sum_{l=1}^{3}\sum_{k=1}^{3} \Pi_{\alpha,kl} \frac{\partial v_{\alpha,l}}{\partial x_k} + H_\alpha\right\}$$

Often you will see this equation with the coefficients 5/3 and 2/3 replaced with γ_α and $\gamma_\alpha - 1$. You do not have to worry about these coefficients. When calculating the internal energy of the particles we have only allowed for 3 degrees of translational freedom. For particles which have more or less than 3 degrees of freedom (e.g. due to rotation and/or vibration) 3 is replaced by $f =$ degrees of freedom, 5/3 is replaced with $(f+2)/f = \gamma$ and 2/3 with $\gamma - 1$. The equation of energy is then

$$\frac{d_\alpha p_\alpha}{dt} - \gamma_\alpha \frac{p_\alpha}{\rho_\alpha}\frac{d_\alpha \rho_\alpha}{dt} = (\gamma_\alpha - 1)\left\{-\nabla . \vec{\Phi}_\alpha - \sum_{l=1}^{3}\sum_{k=1}^{3} \Pi_{\alpha,kl} \frac{\partial v_{\alpha,l}}{\partial x_k} + H_\alpha\right\} \qquad (3.13)$$

This equation is best known in its isentropic version with

$$\Pi_{\alpha,kl} = 0, \quad \vec{\Phi}_\alpha = 0, \quad H_\alpha = 0$$

so that its right hand side is identically zero. Note that the equation of momentum, energy and internal energy are not independent; only two out of three are independent.

Recapitulation

This is a good place to recapitulate and list the macroscopic quantities that we have defined for the particles of type α so far and the equations that we have derived for these macroscopic quantities:

n_α	number of particles per unit volume
ρ_α	mass density
$\vec{v}_\alpha$	average velocity
Q_α	electric charge density per unit volume
$\vec{j}_\alpha$	electric current density
$P_{\alpha,ij}$	pressure tensor
p_α	isotropic pressure
$\Pi_{\alpha,ij}$	stress tensor
T_α	temperature
$\mathcal{U}_\alpha$	internal energy per unit volume
$\vec{\mu}_\alpha$	transfer of momentum due to collisions
$\vec{\Phi}_\alpha$	flux of heat due to random motions
ν_α	transfer of energy due to collisions
H_α	change in internal energy due to collisions

From the previous Sections the reader can select a set of three independent multi-fluid equations for mass, momentum and energy. A possible choice is

mass

$$\frac{\partial n_\alpha}{\partial t} + \nabla(n_\alpha \vec{v}_\alpha) = 0$$

momentum

$$\rho_\alpha \frac{d_\alpha \vec{v}_\alpha}{dt} + \nabla.(p_\alpha \mathbf{I} + \mathbf{\Pi}_\alpha) - \rho_\alpha \vec{g} - Q_\alpha \vec{E} - \vec{j}_\alpha \times \vec{B} = \vec{\mu}_\alpha$$

energy

$$\frac{d_\alpha p_\alpha}{dt} - \gamma_\alpha \frac{p_\alpha}{\rho_\alpha}\frac{d_\alpha \rho_\alpha}{dt} = (\gamma_\alpha - 1)\left\{-\nabla.\vec{\Phi}_\alpha - \sum_{l=1}^{3}\sum_{k=1}^{3}\Pi_{\alpha,kl}\frac{\partial v_l}{\partial x_k} + H_\alpha\right\}$$

These equations are exact consequences of the Boltzmann equation (2.9). They contain information about global quantities related to the particles of type α. We have lost the information on the distribution functions in velocity space. The high-frequency dynamics of the electrons and the low-frequency dynamics of the ions are present in the multi-fluid equations. At the moment the multi-fluid equations are not very useful since as yet the set is not closed: there are more unknowns than equations! This is inherent at the method of moments. In addition there are terms due to the collisions. We shall try to avoid the effects of collisions as much as possible and as a matter of fact we shall be rather successful in that respect. The only occasion where we cannot avoid collisions is when we look at electrical resistivity which is related to collisions of electrons with ions.

3.2 Two-fluid theory

Two-fluid equations for mass, momentum and energy

Plasmas consist of electrons with mass m_e and electric charge $q_e = -e$, ions with mass m_i and electric charge $q_i = Ze$, and possibly neutrals. Here we assume that there are not any neutrals and that there is only one ion component present. If these ions are protons then $Z = 1$. The equations for mass, momentum and energy for such a plasma are easily obtained from (3.14) by taking $\alpha = e, i$:

$$\frac{\partial n_e}{\partial t} + \nabla.(n_e \vec{v}_e) = 0$$

$$\rho_e \frac{d_e v_{e,k}}{dt} + \sum_{l=1}^{3} \frac{\partial}{\partial x_l}(p_e \delta_{kl} + \Pi_{e,kl}) - \rho_e g_k - Q_e E_k - \left(\vec{j}_e \times \vec{B}\right)_k = \mu_{e,k}$$

$$\frac{d_e p_e}{dt} - \gamma_e \frac{p_e}{\rho_e} \frac{d_e \rho_e}{dt} = (\gamma_e - 1)\left\{-\nabla.\vec{\Phi}_e - \sum_{l=1}^{3}\sum_{k=1}^{3} \Pi_{e,kl} \frac{\partial v_{e,l}}{\partial x_k} + H_e\right\}$$

$$Q_e = -en_e,\ \vec{j}_e = Q_e \vec{v}_e = -en_e \vec{v}_e,\ \rho_e = n_e m_e$$

$$\frac{\partial n_i}{\partial t} + \nabla.(n_i \vec{v}_i) = 0$$

$$\rho_i \frac{d_i v_{i,k}}{dt} + \sum_{l=1}^{3} \frac{\partial}{\partial x_l}(p_i \delta_{kl} + \Pi_{i,kl}) - \rho_i g_k - Q_i E_k - \left(\vec{j}_i \times \vec{B}\right)_k = \mu_{i,k}$$

$$\frac{d_i p_i}{dt} - \gamma_i \frac{p_i}{\rho_i} \frac{d_i \rho_i}{dt} = (\gamma_i - 1)\left\{-\nabla.\vec{\Phi}_i - \sum_{l=1}^{3}\sum_{k=1}^{3} \Pi_{i,kl} \frac{\partial v_{i,l}}{\partial x_k} + H_i\right\}$$

$$Q_i = en_i,\ \vec{j}_i = Q_i \vec{v}_i = en_i \vec{v}_i,\ \rho_i = n_i m_i \tag{3.14}$$

$$\frac{d_e}{dt} = \frac{\partial}{\partial t} + \vec{v}_e.\nabla,\quad \frac{d_i}{dt} = \frac{\partial}{\partial t} + \vec{v}_i.\nabla$$

are the total time derivatives with respect to the average flow of the electron fluid and the ion fluid respectively. The equations (3.14) suffer from the same disease as the multi-fluid equations (3.14). They do not form a closed set. We need additional assumptions to obtain closure. We shall come back to this issue later. The two-fluid equations have to be supplemented with Maxwell's equations (2.7) for the electric and magnetic fields. The electromagnetic force couples the electron fluid and the ion fluid as argued in the previous Section for a multi-fluid plasma. A simplified version of this set of two-fluid equations has been used in Chapter II to study the plasma oscillations. The equations used there are obtained by putting

$$p_e = 0,\quad \Pi_{e,kl} = 0$$

$$\vec{B} = 0,\quad \vec{g} = 0$$

$$\vec{\mu}_e = 0, \quad \vec{\Phi}_e = 0, \quad H_e = 0$$

$$p_i = 0, \quad \Pi_{i,kl} = 0$$

$$\vec{\mu}_i = 0, \quad \vec{\Phi}_i = 0, \quad H_i = 0$$

The plasma was free from random motions, it was unmagnetized and there were not any effects due to collisions. In first instance the massive ions were treated as immobile.

Langmuir waves

The two-fluid equations are powerful equations as they contain information ranging from high-frequency short scale dynamics characterized by $\omega_{pe}, \omega_{ce}, \lambda_D, r_{L,e}, \omega_{pi}, \omega_{ci}, r_{L,i}$ up to low-frequency large-scale fluid dynamics. In order to illustrate the potential of the two-fluid equations we consider the temperature correction on the plasma electron oscillations. We relax the assumption that there are not any random motions in the plasma and allow p_e to differ from zero. The two-fluid equations (3.14) now reduce to

$$\frac{\partial n_e}{\partial t} + \nabla.(n_e\vec{v}_e) = 0$$

$$\rho_e\frac{d_e\vec{v}_e}{dt} + \nabla p_e - Q_e\vec{E} = 0$$

$$\frac{d_e p_e}{dt} - \gamma_e\frac{p_e}{\rho_e}\frac{d_e\rho_e}{dt} = 0$$

$$Q_e = -en_e, \quad \vec{j}_e = Q_e\vec{v}_e = -en_e\vec{v}_e, \quad \rho_e = n_e m_e$$

$$\frac{\partial n_i}{\partial t} + \nabla.(n_i\vec{v}_i) = 0$$

$$\rho_i\frac{d_i\vec{v}_i}{dt} + \nabla p_i - Q_i\vec{E} = 0$$

$$\frac{d_i p_i}{dt} - \gamma_i\frac{p_i}{\rho_i}\frac{d_i\rho_i}{dt} = 0$$

$$Q_i = en_i, \quad \vec{j}_i = Q_i\vec{v}_i = en_i\vec{v}_i, \quad \rho_i = n_i m_i \qquad (3.15)$$

To this set (3.15) we must add the relevant Maxwell equation (2.7) for the electric field

$$\varepsilon_0\nabla.\vec{E} = Q_i + Q_e$$

At this point it is instructive to compare (3.15) with the corresponding set of equations (2.18). Where are the additional terms and why are they there? As in Chapter II we start with a plasma where the electrons and ions are initially uniformly distributed, so that the plasma is electrically neutral everywhere. The equations for this uniform static background are:

$$\vec{v}_{e,0} = 0, \quad \vec{v}_{i,0} = 0$$

$$(3.16)$$

$$n_{e,0} = n_{i,0} = \text{constant} \tag{3.17}$$

$$p_{e,0} = \text{constant}, \quad p_{i,0} = \text{constant} \tag{3.18}$$

$$Q_{e,0} = -n_{e,0}e = -n_{i,0}e = -Q_{i,0} \tag{3.19}$$

$$\vec{E}_0 = 0. \tag{3.20}$$

The subscript 0 refers to equilibrium quantities. This system is now perturbed by displacing the electron and ion fluids from their equilibrium positions. The equations for the linear motions of the electron and ion fluids are obtained by linearizing the original equations in the same way as was done in Chapter II. Again we start by treating the heavy ions as immobile in comparison to the light electrons and set

$$\vec{v}_{i,1} = 0, \quad n_{i,1} = 0, \quad Q_{i,1} = 0$$

This leads to the following set of linear equations

$$\frac{\partial n_{e,1}}{\partial t} + n_{e,0}\nabla.\vec{v}_{e,1} = 0$$

$$m_e n_{e,0}\frac{\partial \vec{v}_{e,1}}{\partial t} = -\nabla p_{e,1} - en_{e,0}\vec{E}_1$$

$$\frac{\partial p_{e,1}}{\partial t} - \gamma_e \frac{p_{e,0}}{\rho_{e,0}}\frac{\partial \rho_{e,1}}{\partial t} = 0$$

$$\varepsilon_0 \nabla.\vec{E}_1 = -en_{e,1} \tag{3.21}$$

This set of equations is to be compared with the corresponding set (2.21) for a cold plasma. Since the plasma is no longer cold, (3.21) has an additional electron pressure term in the equation of motion and an equation for internal energy of the electrons. The equations (3.21) have constant coefficients and hence allow solutions in terms of plane waves. This boils down to writing any perturbed quantity $f_1(\vec{x}, t)$ as

$$f_1(\vec{x}, t) = \hat{f}_1 \exp(i(\vec{k}.\vec{x} - \omega t))$$

Here $\vec{k}$ is the wave vector, ω is the frequency and $\hat{f}_1$ is the constant amplitude. The system of linear equations can then be reduced by simple manipulation

$$\frac{\partial}{\partial t} \longrightarrow -i\omega, \quad \nabla \longrightarrow i\vec{k}$$

to a homogeneous system of linear equations. The condition that there are non-trivial solutions to this homogeneous linear system then leads to the dispersion relation that relates the frequency to the wave vector. Exploring this route is left as an exercise for the reader. Another way for solving this system of differential equations (3.21) is to derive an equation for the perturbation of the number density of the electrons $n_{e,1}$

$$\left\{\frac{\partial^2}{\partial t^2} + \omega_{p,e}^2 - \gamma_e v_{t,e}^2 \nabla^2\right\} n_{e,1} = 0 \tag{3.22}$$

This equation is to be compared with (2.22). It has an additional term which transforms it from the linear oscillator equation (2.22) into a wave equation. It is the equation for the *Langmuir waves*. The dispersion relation that corresponds to this equation is known as the "Bohm-Gross dispersion relation"

$$\omega^2 = \omega_{p,e}^2 + k^2\gamma_e v_{t,e}^2 = \omega_{p,e}^2(1 + k^2\gamma_e\lambda_D^2) \qquad (3.23)$$

Compared to (2.23) the important difference is that (3.23) has dispersion, the frequencies depend on the wave vector. From the dispersion relation it follows that there are no Langmuir waves with $\omega < \omega_{p,e}$. Furthermore waves with $\omega > \omega_{p,e}$ only occur as a result of the finite temperature of the electrons. At low k ($k \to 0$), i.e. at long wavelength, or at low temperature ($T_e \to 0$), the phase velocity of the wave ω/k can become arbitrarily large ($\omega/k \to \infty$), and the group velocity $\nabla_k\omega$ goes to zero ($\nabla_k\omega \to 0$), so that no information or energy propagates. This non-propagating wiggle at low k is the plasma oscillation. At high k ($k \to \infty$), i.e. at short wavelength, or at high temperature ($T_e \to \infty$) the Langmuir wave resembles an electron sound wave.

The Langmuir waves are one example of the wide range of phenomena that can be studied with the use of the two-fluid equations. Hence, the two-fluid equations are a powerful tool for studying plasmas. However, the two-fluid equations have their obvious limitations. By integrating over velocity space we have lost the microphysics contained in the distribution function in velocity space. Interaction of waves with particles is out of the scope of the two-fluid equations. For example, Landau damping of Langmuir waves in a collisionless plasma where there is transfer of energy from the waves to particles that have the same velocity as the phase velocity of the wave, requires a fully kinetic treatment using the distribution function in velocity space.

3.3 Single-fluid equations

Going from a multi-fluid description to a single-fluid description is a big step. In single-fluid theory we forget that the plasma is composed of various individual species and use quantities for the plasma as a whole. Hence, by doing so we loose the information on the high-frequency and small-scale dynamics of the electrons and ions characterized by $\omega_{pe}, \omega_{ce}, \lambda_D, r_{L,e}, \omega_{pi}, \omega_{ci}, r_{L,i}$. The single-fluid equations describe the very-low-frequency and large-scale fluid-like behaviour of plasmas. The single-fluid quantities that we are interested in are: mass density ρ, average velocity $\vec{v}$, charge density Q, electrical current density $\vec{j}$, random velocity $\vec{u}$, the stress tensor P_{ij}, pressure p, internal energy $\mathcal{U}$, temperature T, and the flux of heat $\vec{\Phi}$. Mass density ρ, average velocity $\vec{v}$, charge density Q, and electrical current density $\vec{j}$ have already been defined in Chapter II. From equations (2.2), (2.3) and (2.6) we recall that:

$$\begin{aligned} n &= \sum_\alpha n_\alpha, \quad \rho = \sum_\alpha \rho_\alpha = \sum_\alpha n_\alpha m_\alpha \\ \rho\vec{v} &= \sum_\alpha \rho_\alpha \vec{v}_\alpha \\ Q &= \sum_\alpha Q_\alpha, \quad \vec{j} = \sum_\alpha \vec{j}_\alpha \end{aligned} \qquad (3.24)$$

The random velocity $\vec{u}$, the stress tensor P_{ij}, pressure p, internal energy $\mathcal{U}$, temperature T, and the flux of heat $\vec{\Phi}$ are defined as

$$
\begin{aligned}
\vec{u} &= \vec{w} - \vec{v} \\
P_{ij} &= \sum_\alpha m_\alpha \int f_\alpha u_i u_j d^3\vec{w} = \sum_\alpha m_\alpha \int f_\alpha (w_i - v_i)(w_j - v_j) d^3\vec{w} \\
p &= \frac{1}{3}\sum_{i=1}^{3} P_{ii} = \frac{1}{3}\sum_\alpha m_\alpha \int f_\alpha u^2 d^3\vec{w} = \frac{1}{3}\sum_\alpha m_\alpha \int f_\alpha \mid \vec{w} - \vec{v} \mid^2 d^3\vec{w} \\
\Pi_{ij} &= P_{ij} - p\delta_{ij} \\
\rho\mathcal{U} &= \frac{1}{2}\sum_\alpha m_\alpha \int f_\alpha u^2 d^3\vec{w} = \frac{1}{2}\sum_\alpha m_\alpha \int f_\alpha \mid \vec{w} - \vec{v} \mid^2 d^3\vec{w} \\
nk_BT &= \frac{1}{3}\sum_\alpha m_\alpha \int f_\alpha u^2 d^3\vec{w} = p \\
\vec{\Phi} &= \sum_\alpha \frac{m_\alpha}{2} \int f_\alpha u^2 \vec{u} d^3\vec{w} = \sum_\alpha \frac{m_\alpha}{2} \int f_\alpha \mid \vec{w} - \vec{v} \mid^2 (\vec{w} - \vec{v}) d^3\vec{w} \qquad (3.25)
\end{aligned}
$$

Note that the random velocities are now defined with respect to the macroscopic velocity of the plasma as a whole and that the thermodynamic quantities are defined by means of these random velocities. In general

$$p \neq \sum_\alpha p_\alpha, \quad P_{ij} \neq \sum_\alpha P_{\alpha,ij}, \quad \rho\mathcal{U} \neq \sum_\alpha \rho_\alpha \mathcal{U}_\alpha$$

A straightforward calculation (see Problem 13) shows that

$$
\begin{aligned}
p &= \sum_\alpha p_\alpha + \frac{1}{3}\sum_\alpha \rho_\alpha \mid \vec{v}_\alpha - \vec{v} \mid^2 \\
P_{ij} &= \sum_\alpha P_{\alpha,ij} + \sum_\alpha \rho_\alpha (v_{\alpha,i} - v_i)(v_{\alpha,j} - v_j) \\
T &= \frac{1}{n}\sum_\alpha n_\alpha \{T_\alpha + \frac{m_\alpha(\vec{v} - \vec{v}_\alpha)^2}{3k_B}\} \\
p + \frac{1}{3}\rho v^2 &= \sum_\alpha p_\alpha + \frac{1}{3}\sum_\alpha \rho_\alpha v_\alpha^2 \\
\rho\mathcal{U} + \frac{\rho v^2}{2} &= \sum_\alpha (\rho_\alpha \mathcal{U}_\alpha + \frac{\rho_\alpha v_\alpha^2}{2}) \\
P_{ij} + \rho v_i v_j &= \sum_\alpha (P_{\alpha,ij} + \rho_\alpha v_{\alpha,i} v_{\alpha,j}) \\
\Phi_k + (\rho\mathcal{U} + \frac{\rho v^2}{2})v_k + \sum_{l=1}^{3} P_{lk} v_l &= \sum_\alpha \{\Phi_{\alpha,k} + (\rho_\alpha \mathcal{U}_\alpha + \frac{\rho_\alpha v_\alpha^2}{2})v_{\alpha,k} + \sum_{l=1}^{3} P_{\alpha,lk} v_{\alpha,l}\}
\end{aligned}
$$

(3.26)

The first two equalities tell us that total pressure and the total pressure tensor differ from $\sum_\alpha p_\alpha$ and $\sum_\alpha P_{\alpha,ij}$ by quadratic terms in $\vec{v}-\vec{v}_\alpha$. A similar result holds for the temperatures. According to the third equality the temperature of the plasma as a whole T differs from the mean of the temperatures of the different plasma components $(\sum_\alpha n_\alpha T_\alpha)/n$ by quadratic terms in $\vec{v}-\vec{v}_\alpha$.

The equations for density ρ, velocity $\vec{v}$ and total energy (and hence for pressure p and internal energy) are obtained by summing the corresponding multi-fluid equations over the various species:

$$\sum_\alpha (\text{conservation of mass for species } \alpha)$$

$$\sum_\alpha (\text{conservation of momentum for species } \alpha)$$

$$\sum_\alpha (\text{conservation of energy for species } \alpha)$$

For what follows it is helpful to note that the operators $\sum_\alpha$, $\frac{\partial}{\partial t}$ and $\frac{\partial}{\partial x_k}$ can be interchanged.

Conservation of mass

We start from the equation of mass for particles of species α (3.2) and sum over all species α. With the use of (3.24) we obtain the equation of total mass conservation

$$\boxed{\frac{\partial \rho}{\partial t} + \nabla.(\rho\vec{v}) = 0 \qquad (3.27)}$$

Note that this equation is written as a conservation law. At this point it is convenient to introduce the total time derivative or convective time derivative or Lagrangian time derivative with respect to the flow of the plasma as a whole:

$$\boxed{\frac{d}{dt} = \frac{\partial}{\partial t} + \vec{v}.\nabla \qquad (3.28)}$$

With this total time derivative (3.28) the equation of total mass conservation is

$$\boxed{\frac{d\rho}{dt} + \rho\nabla.\vec{v} = 0 \qquad (3.29)}$$

Conservation of charge

We multiply the equation for conservation of number density of particles of type α (3.2) by q_α and add over all species α to obtain the equation for conservation of charge:

$$\boxed{\frac{\partial Q}{\partial t} + \nabla.\vec{j} = 0 \qquad (3.30)}$$

Conservation of momentum

We start from the equation of momentum for particles of species α (3.7) and sum over all α. With the use of (3.24), (3.25) and (3.43), and taking into account that $\sum_\alpha \vec{\mu}_\alpha = 0$ we obtain the equation of motion for the plasma as a whole

$$\frac{\partial}{\partial t}(\rho\vec{v}) + \nabla.(\rho\vec{v}\vec{v} + \mathbf{P}) - \rho\vec{g} - Q\vec{E} - \vec{j}\times\vec{B} = 0 \tag{3.31}$$

This is the momentum conservation equation in Magnetohydrodynamics. With the use the total time derivative (3.28) we can rewrite the equation of motion as

$$\rho\frac{d\vec{v}}{dt} = -\nabla.\mathbf{P} + \rho\vec{g} + Q\vec{E} + \vec{j}\times\vec{B} \tag{3.32}$$

We can identify the force due to the pressure tensor $-\nabla.\mathbf{P}$, the gravitational force $\rho\vec{g}$, the electric force $Q\vec{E}$ and the Lorentz force $\vec{j}\times\vec{B}$.

Interludium 1

What now follows is an interludium which can be skipped on a first reading. Its aim is to write the equation of motion in the form of a conservation law. The first form of the equation of motion is almost in the form of a conservation law. In order to cast (3.31) in the form of a conservation law we have to rewrite $-\rho\vec{g} - Q\vec{E} - \vec{j}\times\vec{B}$ in terms of partial derivatives with respect to time t and x_k. This is easier said than done. Let us have a look at the electromagnetic force and see if we can write it in conserved form. Of course we know we can. We have learned about Maxwell's stress tensor and the momentum associated with the electromagnetic field. In Problem 14 you are asked to show that the electromagnetic force can be written as

$$Q\vec{E} + \vec{j}\times\vec{B} = \nabla.\mathbf{M} - \frac{\partial}{\partial t}\left(\epsilon_0\vec{E}\times\vec{B}\right)$$

$\mathbf{M}$ is Maxwell's stress tensor

$$\mathbf{M} = -\left(\frac{\epsilon_0 E^2}{2} + \frac{B^2}{2\mu_0}\right)\mathbf{I} + \epsilon_0\vec{E}\vec{E} + \frac{\vec{B}\vec{B}}{\mu_0}$$

and

$$\frac{\epsilon_0 E^2}{2} + \frac{B^2}{2\mu_0}$$

is the isotropic pressure due to the electromagnetic field. Now let us have a look at the gravitational force. Write $\vec{g}$ as $\vec{g} = -\nabla\Psi$ with Ψ the gravitational potential. In case $\vec{g}$ is caused by self-gravitation we have $\nabla^2\Psi = 4\pi G\rho$, $\nabla.\vec{g} = -4\pi G\rho$. A straightforward calculation shows that

$$\rho g_i = \sum_{k=1}^{3}\frac{\partial T_{ik}}{\partial x_k}, \quad T_{ik} = \frac{-1}{4\pi G}(g_i g_k - \frac{g^2\delta_{ik}}{2})$$

where T_{ik} is the gravitational stress tensor. We can combine these results to write the equation of motion in the form of a conservation law as

$$\begin{aligned}\frac{\partial}{\partial t}\left(\rho\vec{v} + \epsilon_0(\vec{E}\times\vec{B})\right) + \nabla.\{\rho\vec{v}\vec{v} + \mathbf{P} - \mathbf{M} - \mathbf{T}\} &= 0 \\ \frac{\partial}{\partial t}\left(\rho\vec{v} + \epsilon_0(\vec{E}\times\vec{B})\right) + \nabla.\mathbf{A} &= 0 \qquad (3.33)\end{aligned}$$

This equation is written in the form of a conservation law. The operator $\frac{\partial}{\partial t}$ acts on terms that correspond to momentum: ρv_i is the momentum associated with the motion of the plasma as a whole, $\epsilon_0 \vec{E} \times \vec{B}$ can be seen as the momentum associated with the electromagnetic field. The operator $\nabla = \sum_{k=1}^{3} \frac{\partial}{\partial x_k}$ acts on terms that correspond stresses. Already known are the pressure tensor **P**, the Maxwell stress tensor **M** and the gravitational stress tensor **T**. $\rho\vec{v}\vec{v}$ is called the stress tensor of Reynolds. It is the macroscopic twin of the viscosity stress tensor which is buried in **P**. $\mathbf{A} = \mathbf{P} + \rho\vec{v}\vec{v} - \mathbf{M} - \mathbf{T}$ is the so-called grand momentum stress tensor.

Conservation of energy

We start from the equation of energy for particles of species α (3.11) and add over all types α. With the use of (3.24), (3.25) and (3.43), and taking into account that $\sum_\alpha \nu_\alpha = 0$ we obtain the equation for conservation of energy for the whole plasma

$$\frac{\partial}{\partial t}\left(\frac{\rho v^2}{2} + \rho\mathcal{U}\right) + \nabla.\left\{\left(\frac{\rho v^2}{2} + \rho\mathcal{U}\right)\vec{v} + \vec{\Phi}\right\} + \sum_{k=1}^{3}\sum_{l=1}^{3}\frac{\partial}{\partial x_k}(v_l P_{kl}) - \rho\vec{v}.\vec{g} - \vec{j}.\vec{E} = 0 \tag{3.34}$$

With the use of the total time derivative we can rewrite the equation of energy as

$$\rho\frac{d}{dt}\left(\frac{v^2}{2} + \mathcal{U}\right) + \nabla.\vec{\Phi} + \sum_{k=1}^{3}\sum_{l=1}^{3}\frac{\partial}{\partial x_k}(v_l P_{kl}) - \rho\vec{v}.\vec{g} - \vec{j}.\vec{E} = 0 \tag{3.35}$$

We can identify the work done by the stress tensor $-\sum_{k=1}^{3}\frac{\partial}{\partial x_k}\left(\sum_{l=1}^{3} v_l P_{kl}\right)$, the work done by the gravitational force $\rho\vec{v}.\vec{g}$, and the work done by the electromagnetic forces $\vec{j}.\vec{E}$ (see problem 15).

Interludium 2

Here is another interludium. As the first one it can be skipped on a first reading. Its aim is to write the equation of energy in the form of a conservation law. The first form of the equation of energy is almost in the form of a conservation law. We have to rewrite

$$\rho\vec{v}.\vec{g} + \vec{j}.\vec{E}$$

in terms of partial derivatives with respect to time t and x_k. This is easier said than done. Let us have a look at the work done by the electromagnetic forces and see if we can write it in conserved form. Of course we know we can. We have learned about *Poynting's theorem*: (see also Problem 16)

$$\vec{j}.\vec{E} = -\nabla.\left(\frac{\vec{E}\times\vec{B}}{\mu_0}\right) - \frac{\partial}{\partial t}\left(\frac{\epsilon_0 E^2}{2} + \frac{B^2}{2\mu_0}\right)$$

In Problem 15 you are asked to verify that $\vec{j}.\vec{E}$ is the work done per unit time by the electromagnetic forces. Poynting's theorem is telling us that the energy density of the electromagnetic field

$$\frac{\epsilon_0 E^2}{2} + \frac{B^2}{2\mu_0}$$

changes in time because of the flux of energy of the electromagnetic field, known as the *Poynting flux*

$$\frac{\vec{E}\times\vec{B}}{\mu_0}$$

and because of the work done per unit time by the electromagnetic forces. OK, we are happy with Poynting's theorem as it provides us with an expression for $\vec{j}.\vec{E}$ in the form of a conservation law.

Let us have a look at the work done by the gravitational force and see whether it can be written in the form of a conservation law. Straightforward calculation shows that (see problem 17)

$$-\rho\vec{v}.\vec{g} = \nabla.(\rho\Psi\vec{v}) + \frac{\partial}{\partial t}(\rho\Psi) - \rho\frac{\partial\Psi}{\partial t}$$

You cannot always get what you want. We do not like the term $-\rho\frac{\partial\Psi}{\partial t}$ but we cannot do any better! $\rho\vec{v}.\vec{g}$ is the work done by the gravitational force. It causes a change in the gravitational energy $\rho\Psi$, a flux in gravitational energy $\rho\Psi\vec{v}$ and the term $-\rho\frac{\partial\Psi}{\partial t}$. Combining these results we can write the equation of energy as

$$\frac{\partial}{\partial t}\{\frac{\rho v^2}{2} + \rho\mathcal{U} + \frac{\epsilon_0 E^2}{2} + \frac{B^2}{2\mu_0} + \rho\Psi\} +$$

$$\nabla.\{(\frac{\rho v^2}{2} + \rho\mathcal{U} + \rho\Psi)\vec{v} + \frac{\vec{E}\times\vec{B}}{\mu_0} + \vec{\Phi}\} + \sum_{k=1}^{3}\sum_{l=1}^{3}\frac{\partial}{\partial x_k}(v_l P_{kl}) = \rho\frac{\partial\Psi}{\partial t} \tag{3.36}$$

The operator $\partial/\partial t$ acts on the different forms of energy density: kinetic energy, internal energy, electromagnetic energy and potential energy. The operator ∇ acts on terms that correspond to the different ways in which energy can enter or leave a volume of plasma through its bounding surface: $(\rho v^2/2 + \rho\mathcal{U} + \rho\Psi)\vec{v}$ is the flux of kinetic energy + internal energy + gravitational energy due to the global motion of the plasma; $\sum_{k=1}^{3}\partial/\partial x_k(\sum_{l=1}^{3} v_l P_{kl})$ is the work done by the stress tensor due to the global motion; $(\vec{E}\times\vec{B})/\mu_0$ is the Poynting flux or the flux of electromagnetic energy and Φ is the flux of energy due to the random motions in the plasma. The latter flux has the effect that there can be a flux of energy present in a plasma without a global motion.

Conservation of internal energy

We take the scalar product of the equation for the conservation of momentum (3.32) with the average velocity $\vec{v}$ and find the equation for the change in kinetic energy as

$$\rho\frac{d}{dt}\left(\frac{v^2}{2}\right) + \sum_{l=1}^{3}\sum_{k=1}^{3}\frac{\partial P_{kl}}{\partial x_k}v_l - \rho\vec{g}.\vec{v} - Q\vec{v}.\vec{E} - (\vec{j}\times\vec{B}).\vec{v} = 0$$

When we subtract this equation from the equation for the change in total energy (3.35) we find the equation for the change in internal energy

$$\boxed{\rho\frac{d\mathcal{U}}{dt} + \sum_{l=1}^{3}\sum_{k=1}^{3}P_{kl}\frac{\partial v_l}{\partial x_k} + \nabla.\vec{\Phi} - \vec{E}.(\vec{j} - Q\vec{v}) + (\vec{j}\times\vec{B}).\vec{v} = 0}$$

The terms due to the electromagnetic field can be written as (see Problem 18)

$$-\vec{E}.(\vec{j} - Q\vec{v}) + (\vec{j} \times \vec{B}).\vec{v} = -\vec{E}^{\star}.\vec{j}^{\star}$$

with

$$\vec{E}^{\star} = \vec{E} + \vec{v} \times \vec{B} \;\wedge\; \vec{j}^{\star} = \vec{j} - Q\vec{v}$$

We have new quantities: $\vec{E}^{\star}$ and $\vec{j}^{\star}$. Do they mean more problems or extra fun? Recall that

$$\vec{j} = \sum_{\alpha} \vec{j}_{\alpha} = \sum_{\alpha} n_{\alpha} q_{\alpha} \vec{v}_{\alpha} = \sum_{\alpha} n_{\alpha} q_{\alpha} (\vec{v}_{\alpha} - \vec{v}) + Q\vec{v}$$

so that

$$\boxed{\vec{j}^{\star} = \vec{j} - Q\vec{v} = \sum_{\alpha} n_{\alpha} q_{\alpha} (\vec{v}_{\alpha} - \vec{v}) \qquad (3.37)}$$

$Q\vec{v}$ is the *convective electric current density*, it is caused by the net charge of positive or negative electric charges that are convected by the flow of the plasma as a whole. $\vec{j}^{\star} = \sum_{\alpha} n_{\alpha} q_{\alpha} (\vec{v}_{\alpha} - \vec{v})$ is the *conductive electric current density*, caused by the relative motions of the positive and negative charges. It is the electric current measured in a reference system that moves with the plasma with velocity $\vec{v}$. Go back to the previous Chapter and recall *electrical quasi-neutrality* of a plasma. Realize that electrical quasi-neutrality means that the electrical charge density is zero to high degree of accuracy over spatial scales sufficiently larger than λ_D and times scales sufficiently longer than the electron plasma period : $Q \approx 0$ and consequently

$$\mid Q\vec{v} \mid << \mid \vec{j} \mid, \;\; \vec{j} = \vec{j}^{\star}$$

$\vec{E}^{\star} = \vec{E} + \vec{v} \times \vec{B}$ is the electric field measured in a reference frame that moves with velocity $\vec{v}$; it is the electric field seen by an observer who moves with the plasma. In the following Chapter we shall show that

$$\vec{E}^{\star} \approx 0 \;\text{ in ideal MHD}$$

$$\vec{E}^{\star} \approx \eta\vec{j} \;\text{ in resistive MHD}$$

OK here is a preview, which you can skip if you want. In resistive MHD

$$\boxed{\vec{E}^{\star}.\vec{j}^{\star} = \eta j^2 \qquad (3.38)}$$

is the Ohmic heating. You will hear more about this term in the next Chapter. We use this result to write the equation for internal energy as

$$\rho\frac{d\mathcal{U}}{dt} + \sum_{l=1}^{3}\sum_{k=1}^{3} P_{kl}\frac{\partial v_l}{\partial x_k} = -\nabla.\Phi + \vec{E}^{\star}.\vec{j}^{\star}$$

A simple calculation and use of the equation of mass conservation enables us to rewrite the contribution due to the pressure tensor.

$$\begin{aligned} P_{kl} &= p\delta_{kl} + \Pi_{kl} \\ \sum_{l=1}^{3}\sum_{k=1}^{3} P_{kl}\frac{\partial v_l}{\partial x_k} &= -\frac{p}{\rho}\frac{d\rho}{dt} + \sum_{l=1}^{3}\sum_{k=1}^{3} \Pi_{kl}\frac{\partial v_l}{\partial x_k} \end{aligned}$$

We use this result for the contribution of the pressure tensor and recall that $\rho\mathcal{U} = 3p/2$ to obtain what is probably the most popular version of the equation for internal energy

$$\frac{dp}{dt} - \frac{5}{3}\frac{p}{\rho}\frac{d\rho}{dt} = \frac{2}{3}\left\{-\nabla.\Phi + \vec{E}^{\star}.\vec{j}^{\star} - \sum_{l=1}^{3}\sum_{k=1}^{3} \Pi_{kl}\frac{\partial v_l}{\partial x_k}\right\}$$

Often you will see this equation with the coefficients $5/3$ and $2/3$ replaced with γ and $\gamma - 1$. You do not have to worry about these coefficients. When calculating the internal energy of the particles we have only allowed for 3 degrees of translational freedom. For particles which have more than 3 degrees of freedom (e.g. due to rotation and/or vibration) 3 is replaced by $f =$ degrees of freedom, $5/3$ is replaced with $(f+2)/f = \gamma$ and $2/3$ with $\gamma - 1$. The equation of energy is then

$$\boxed{\frac{dp}{dt} - \frac{\gamma p}{\rho}\frac{d\rho}{dt} = (\gamma - 1)\left\{-\nabla.\Phi + \vec{E}^{\star}.\vec{j}^{\star} - \sum_{l=1}^{3}\sum_{k=1}^{3} \Pi_{kl}\frac{\partial v_l}{\partial x_k}\right\}} \quad (3.39)$$

This is the equation for internal energy in Magnetohydrodynamics. We can write this equation in terms of temperature by recalling that

$$p = nk_BT = \frac{k_B}{\overline{m}}\rho T = \frac{k_B}{m_p\overline{\mu}}\rho T = \frac{\overline{\mathcal{R}}}{\overline{\mu}}\rho T \quad (3.40)$$

$\overline{m}$ is the mean mass of the particles and $\overline{\mu}$ is the mean atomic weight. $\overline{m}$, $\overline{\mu}$ and $\overline{\mathcal{R}}$ are defined as

$$\begin{aligned} \overline{m} &= \sum_{\alpha} n_\alpha m_\alpha / n = \rho/n \\ \overline{m} &= \overline{\mu} m_p \\ \overline{\mathcal{R}} &= \frac{k_B}{m_p} = 8.2548 \times 10^3 \mathrm{m}^2\mathrm{s}^{-2}\mathrm{K}^{-1} \end{aligned}$$

Recapitulation

From the previous subsections we can select different sets of independent single-fluid equations for mass, total charge, momentum and energy. A possible selection is:

mass

$$\frac{d\rho}{dt} + \rho\nabla.\vec{v} = 0$$

electric charge

$$\frac{\partial Q}{\partial t} + \nabla.\vec{j} = 0$$

momentum

$$\rho\frac{d\vec{v}}{dt} + \nabla.\mathbf{P} - \rho\vec{g} - Q\vec{E} - \vec{j}\times\vec{B} = 0$$

internal energy

$$\frac{dp}{dt} - \frac{\gamma p}{\rho}\frac{d\rho}{dt} = (\gamma - 1)\left\{-\nabla.\Phi + \vec{E}^{\star}.\vec{j}^{\star} - \sum_{l=1}^{3}\sum_{k=1}^{3}\Pi_{kl}\frac{\partial v_l}{\partial x_k}\right\} \tag{3.41}$$

Closure

Have a very good look at these equations. They are exact consequences of the Boltzmann equation (2.9). They are non-linear partial differential equations (why?). Note that we have kept our promise; you do not see any term that contains the collision term. Collisional invariance has done a very good job. That is the good news, now the bad news. The equations do not form a closed set. You know the reason. We have taken the zeroth moment of the Boltzmann equation (2.9); that gave us an equation for density ρ and 3 new first order unknowns, namely the 3 components of velocity $\vec{v}$. We have taken the 3 scalar first order moments of the Boltzmann equation (2.9) (multiplication with w_i) and that gave us 3 scalar equations for the components of velocity (or 1 vector equation for the velocity) but also 6 new second order unknowns, namely the elements of the stress tensor P_{ij}, $i = 1,2,3; j \geq i$. We have taken 1 second order moment of the Boltzmann equation (2.9) (multiplication with w^2) and that gave us 1 equation for pressure (= trace of the stress tensor) and 3 additional third order unknowns, namely the 3 components of the flux of heat due to random motions Φ. Equations for all elements of the stress tensor could be obtained by taking 5 additional second order moments of the Boltzmann equation (2.9) (which 5?), but that would give us additional third order unknowns. The conclusion is that we do not have any equations for the anisotropic part of the stress tensor and for the flux of heat due to random motions. This is not unexpected but inherent at the method of moments.

However it is more complicated than that. Expressions for the anisotropic part of the stress tensor and for the flux of heat due to random motions do not suffice for obtaining a closed set of equations. In order to make that point clear, let us look at a stationary situation ($\partial/\partial t = 0$) which has a minimal number of unknowns but is still a MHD problem. We forget about the anisotropic part of the stress tensor and the flux of heat for the moment. The unknowns in this situation are $\vec{v}, p, \rho, \vec{B}$. The electric current density $\vec{j}$ can be calculated by using one of Maxwell's equations (2.7). The relevant equations are the equation of mass conservation, the equation of motion, the equation of internal energy and the remaining Maxwell equation (2.7)

$$\nabla.(\rho\vec{v}) = 0$$

$$\rho(\vec{v}.\nabla)\vec{v} = -\nabla p + \frac{1}{\mu}(\nabla \times \vec{B}) \times \vec{B}$$

$$\vec{v}.\nabla p = \frac{\gamma p}{\rho}\vec{v}.\nabla \rho$$

$$\nabla.\vec{B} = 0 \tag{3.42}$$

Hence there are 8 unknowns (3 components of velocity, density, pressure and the 3 components of the magnetic field) and we have 6 scalar equations in total.

The message is very clear now. Expressions for the anisotropic part of the stress tensor and for the flux of heat due to random motions are a necessary but not a sufficient condition for closure. We need an additional equation to close our system of single-fluid equations. In order to understand why we need an additional equation, we have to go back to the equation for conservation of momentum (3.31), (3.32). This equation contains the electrical current density, $\vec{j}$, as a new variable. Hence, to close the system we need an equation that links $\vec{j}$ to the plasma variables $\rho, P_{i,j}, \vec{v}$. This equation will be obtained by computing the first charge moment of the Boltzmann equation (2.9) which is the *generalized Ohm's law.* The generalized Ohm's law and its approximate MHD form are at the heart of MHD and deserve a chapter of their own.

3.4 Recapitulation

"For all your education, Ben, you seem rather naive about certain things. One of them is that someday you are going to have to earn a living."
"Am I?"
"Of course."
"Are you going broke or something? You can't afford to feed me any more?"
Conversation of Benjamin Braddock with his father, The graduate.
C. Webb

- Kinetic theory: Boltzmann-Maxwell equations
  The full set of Boltzmann-Maxwell equations (2.7) - (2.9) provides a very detailed description of plasma behaviour. However, this set of equations is virtually impossible to solve in any nontrivial situation. This realization has led to the development of simpler mathematical models with a narrower range of applicability.

- Multi-fluid theory.
  Multi-fluid theory recognizes that the plasma is composed of different particle species

and assumes that each species behaves as a separate fluid. It captured the differences in the fluid behaviour of the light electrons and the heavy ions.

- Multi-fluid theory.
  Multi-fluid theory derives differential equations that govern the temporal and spatial variations of the macroscopic variables directly from the Bolzmann equation (2.9) without solving it. These differential equations are obtained by taking the moments of the Boltzmann equation (2.9). The first three moments give us the equation of conservation of mass (continuity equation), the equation of conservation of momentum (equation of motion), and the equation of conservation of energy (energy equation) for particles of type α.

- Multi-fluid theory.
  However, at each stage of the hierarchy of moments, the resulting set of equations is not complete. Each time a higher moment of the Bolzmann equation is calculated a new macroscopic variable appears. For example, the continuity equation relates the number density and the mean velocity. The inclusion of the momentum equation adds the kinetic pressure tensor, and the inclusion of the energy equation adds the heat flow vector. It is thus necessary to truncate the system of equations at some stage of hierarchy by introducing a simplifying assumption concerning the highest moment of the distribution function that appears in the system. This truncation creates a closed system of equations.

- Multi-fluid theory.
  The multi-fluid equations are exact consequences of the Boltzmann equation (2.9). They contain information about global quantities related to the particles of type α ranging from high-frequency short scale dynamics up to low-frequency large-scale fluid dynamics. By integrating over velocity space the microphysics related to the distribution functions in velocity space is lost.

- Multi-fluid theory.
  The electromagnetic force couples the equation for conservation of momentum to the full set of Maxwell's equations (2.7) and makes it very distinct from the Navier-Stokes equation of conventional hydrodynamics where we have only pressure and viscous forces acting on the fluid elements. The electromagnetic force also couples all the charged plasma components together.

- Two-fluid theory.
  The two-fluid equations are a special case of the multi-fluid equations for a fully ionized plasma consisting of electrons and one type of ions. They are powerful equations as they contain information ranging from high-frequency short-scale dynamics characterized by $\omega_{pe}, \omega_{ce}, \lambda_D, r_{L,e}, \omega_{pi}, \omega_{ci}, r_{L,i}$ up to low-frequency large-scale fluid dynamics. As an example of their strength they are used to determine the temperature correction on the plasma electron oscillations.

- Single-fluid theory.
  So far the equations are written separately for each particle species. However, plasma can also be considered as a conducting fluid (one fluid theory) without specifying its various individual species. In this case, each macroscopic variable is formed by adding

the contributions of the various particle species in the plasma. The equations for density, velocity and pressure are obtained by summing the corresponding equations for each species and by using collisional invariance. The high-frequency short scale dynamics of the electrons and ions is removed when adopting the single-fluid equations for describing a plasma. Single-fluid theory is restricted to large-scale low-frequency fluid behaviour.

- Single-fluid theory.
  The single-fluid equations for mass, momentum and internal energy do not form a closed system.

3.5 Problems

> "I'll figure it out.
> I'm going to use all the power of my brain."
> *Homer Simpson*
> The Simpsons

1. Go back to the continuity equation for particles of species α (3.2). Compute the terms T_1 and T_2 and find

$$T_1 = \int \frac{\partial f_\alpha}{\partial t} d^3\vec{w} = \frac{\partial n_\alpha}{\partial t}, \quad T_2 = \int \vec{w}.\nabla_x f_\alpha d^3\vec{w} = \nabla_x.(n_\alpha \vec{v}_\alpha)$$

   Show that

$$\frac{\vec{F}_\alpha}{m_\alpha}.\nabla_w f_\alpha = \nabla_w.\left(\frac{f_\alpha}{m_\alpha}\vec{F}_\alpha\right)$$

   and use this result to compute T_3

$$T_3 = \int \nabla_w.\left(\frac{f_\alpha}{m_\alpha}\vec{F}_\alpha\right) d^3\vec{w} = \int \frac{f_\alpha}{m_\alpha}\vec{F}_\alpha.d\vec{S}_\infty = 0$$

2. Go back to the equation for conservation of momentum for particles of species α (3.7). Compute T_1 and T_2

$$\begin{aligned} T_1 &= \int \frac{\partial f_\alpha}{\partial t} m_\alpha w_k d^3\vec{w} = \frac{\partial}{\partial t}(\rho_\alpha v_{\alpha,k}) \\ T_2 &= \int m_\alpha w_k \vec{w}.\nabla_x f_\alpha d^3\vec{w} = \sum_{l=1}^{3} \frac{\partial}{\partial x_l}(\rho_\alpha < w_k w_l >_\alpha) \end{aligned}$$

3. Show that

$$< w_k w_l >_\alpha = v_{\alpha,k} v_{\alpha,l} + < u_{\alpha,k} u_{\alpha,l} >_\alpha$$

4. A system of particles e.g. electrons is characterized by a uniform and time independent distribution function $f(w_x, w_y, w_z)$.
Compute the stress tensor P_{ij}, the scalar isotropic pressure p, the temperature T and the heat flow vector (= the flux of heat due to random motions) for

- the Maxwell-Boltzman distribution function

$$f(w_x, w_y, w_z) = n\left(\frac{m}{2\pi k_B T}\right)^{3/2} \exp\left(-\frac{m(w_x^2 + w_y^2 + w_z^2)}{2k_B T}\right)$$

- the bi-Maxwell-Boltzman distribution function

$$f(w_x, w_y, w_z) = \frac{nm}{2\pi k_B T_\perp}\left(\frac{m}{2\pi k_B T_{||}}\right)^{1/2} \exp\left\{-\frac{m}{2k}\left(\frac{w_x^2 + w_y^2}{T_\perp} + \frac{w_z^2}{T_{||}}\right)\right\}$$

(B1986)

5. Show that

$$w_k \vec{F}_\alpha . \nabla_w f_\alpha = \nabla_w . \left(f_\alpha w_k \vec{F}_\alpha\right) - f_\alpha F_{\alpha,k}$$

and use this result to compute

$$T_3 = \int w_k \vec{F}_\alpha . \nabla_w f_\alpha d^3\vec{w} = -\rho_\alpha g_k - Q_\alpha E_k - \left(\vec{j}_\alpha \times \vec{B}\right)_k$$

6. Use the random velocities to compute

$$\mu_{\alpha,k} = m_\alpha \int C_\alpha u_{\alpha,k} d^3\vec{w} = m_\alpha \sum_\beta \int C_{\alpha,\beta} u_{\alpha,k} d^3\vec{w}$$

Note that

$$\int C_\alpha d^3\vec{w} = 0$$

7. Go back to the equation for conservation of energy for particles of species α (3.11). Compute T_1 and T_2

$$T_1 = \int \frac{m_\alpha w^2}{2}\frac{\partial f_\alpha}{\partial t} d^3\vec{w} = \frac{\partial}{\partial t}\left(\frac{\rho_\alpha v_\alpha^2}{2} + \rho_\alpha \mathcal{U}_\alpha\right)$$

$$T_2 = \int \frac{m_\alpha w^2}{2} \vec{w}.\nabla_x f_\alpha d^3\vec{w} = \nabla.\left(\left(\frac{\rho_\alpha v_\alpha^2}{2} + \rho_\alpha \mathcal{U}_\alpha\right)\vec{v}_\alpha + \frac{\rho_\alpha < u_\alpha^2 \vec{u}_\alpha >_\alpha}{2}\right)$$

$$+ \sum_{k=1}^{3} \frac{\partial}{\partial x_k}\left(\sum_{l=1}^{3} P_{\alpha,kl} v_{\alpha,l}\right)$$

A helpful result for the second equality is

$$w_k w^2 = (v_{\alpha,k} + u_{\alpha,k})(v_\alpha^2 + u_\alpha^2 + 2\sum_{l=1}^{3} v_{\alpha,l} u_{\alpha,l})$$

$$< w_k w^2 >_\alpha = v_\alpha^2 v_{\alpha,k} + < u_\alpha^2 >_\alpha v_{\alpha,k} + < u_\alpha^2 u_{\alpha,k} >_\alpha + 2\sum_{l=1}^{3} v_{\alpha,l} < u_{\alpha,k} u_{\alpha,l} >_\alpha$$

8. Note that

$$\sum_{k=1}^{3} \frac{\partial F_{\alpha,k}}{\partial w_k} = 0, \quad (\vec{w} \times \vec{B}).\vec{w} = 0$$

and show that

$$\frac{w^2}{2}\vec{F}_\alpha.\nabla_w f_\alpha = \nabla_w.\left(\frac{w^2}{2} f_\alpha \vec{F}_\alpha\right) - f_\alpha \vec{F}_\alpha.\vec{w}$$

Compute then

$$T_3 = \int \frac{w^2}{2}\vec{F}_\alpha.\nabla_w f_\alpha d^3\vec{w} = -\int f_\alpha \vec{F}_\alpha.\vec{w} d^3\vec{w} = -\rho_\alpha \vec{v}_\alpha.\vec{g} - \vec{j}_\alpha.\vec{E}$$

9. Use the random velocities to compute ν_α as $\nu_\alpha = H_\alpha + \vec{v}_\alpha.\vec{\mu}_\alpha$. Here also remember that $\int C_\alpha d^3\vec{w} = 0$

10. Use the equation for conservation of mass (3.2) to show that

$$\begin{aligned}
\frac{\partial}{\partial t}(\rho_\alpha v_{\alpha,k}) + \sum_{l-1}^{3} \frac{\partial}{\partial x_l}(\rho_\alpha v_{\alpha,k} v_{\alpha,l}) &= \rho_\alpha \frac{d_\alpha v_{\alpha,k}}{dt} \\
\frac{\partial}{\partial t}(\frac{\rho_\alpha v_\alpha^2}{2} + \rho_\alpha \mathcal{U}_\alpha) + \nabla.\{(\frac{\rho_\alpha v_\alpha^2}{2} + \rho_\alpha \mathcal{U}_\alpha)\vec{v}_\alpha\} &= \rho_\alpha \frac{d_\alpha}{dt}(\frac{v_\alpha^2}{2} + \mathcal{U}_\alpha)
\end{aligned}$$

and then use these results to rewrite the equations for conservation of momentum and of energy as

$$\begin{aligned}
\rho_\alpha \frac{d_\alpha v_{\alpha,k}}{dt} &+ \sum_{l=1}^{3} \frac{\partial}{\partial x_l}(p_\alpha \delta_{kl} + \Pi_{\alpha,kl}) \\
&- \rho_\alpha g_k - Q_\alpha E_k - \left(\vec{j}_\alpha \times \vec{B}\right)_k = \mu_{\alpha,k} \\
\rho_\alpha \frac{d_\alpha}{dt}(\frac{v_\alpha^2}{2} + \mathcal{U}_\alpha) &+ \sum_{k=1}^{3} \frac{\partial}{\partial x_k}(\Phi_{\alpha,k} + \sum_{l=1}^{3} v_{\alpha,l} P_{\alpha,kl}) \\
&- \rho_\alpha \vec{v}_\alpha.\vec{g} - \vec{j}_\alpha.\vec{E} = \nu_\alpha = H_\alpha + \vec{v}_\alpha.\vec{\mu}_\alpha
\end{aligned}$$

11. Go back to the subsection on Langmuir waves. Start from the equations for the linear motions of the electron fluid (the heavy ions are treated as immobile)

$$\frac{\partial n_{e,1}}{\partial t} + n_{e,0}\nabla(\vec{v}_{e,1}) = 0$$

$$m_e n_{e,0} \frac{\partial \vec{v}_{e,1}}{\partial t} = -\nabla p_{e,1} - e\vec{E}_1$$

$$\frac{\partial p_{e,1}}{\partial t} - \gamma_e \frac{p_{e,0}}{\rho_{e,0}} \frac{\partial \rho_{e,1}}{\partial t} = 0$$

$$\varepsilon_0 \nabla \vec{E}_1 = -e n_{e,1}$$

Look for solutions in terms of plane waves and write the linear quantities $n_{e,1}, p_{e,1}, \vec{v}_{e,1}, \vec{E}_1$ as

$$f_1(\vec{x},t) = \hat{f}_1 \exp(i(\vec{k}.\vec{x} - \omega t))$$

Here $\vec{k}$ is the wave vector, ω is the frequency and $\hat{f}_1$ is the constant amplitude. The system of linear ordinary equations can then be reduced by simple manipulation

$$\frac{\partial}{\partial t} \longrightarrow -i\omega \quad \nabla \longrightarrow i\vec{k}$$

to a homogeneous system of linear equations. The condition that there are non-trivial solutions to this homogeneous linear system then leads to the dispersion relation that relates the frequency to the wave vector.

12. Derive the equation for the perturbation of the number density of the electrons $n_{e,1}$ in a finite temperature non-magnetic plasma when the ions are immobile:

$$\left\{ \frac{\partial^2}{\partial t^2} + \omega_{p,e}^2 - \gamma_e v_{t,e}^2 \nabla^2 \right\} n_{e,1} = 0$$

13. Go back to the Section on single-fluid theory. Consider a plasma which is made up of a mixture of particles of various species α. Show that:

$$\begin{aligned}
p &= \sum_\alpha p_\alpha + \frac{1}{3}\sum_\alpha \rho_\alpha \mid \vec{v}_\alpha - \vec{v} \mid^2 \\
P_{ij} &= \sum_\alpha P_{\alpha,ij} + \sum_\alpha \rho_\alpha (v_{\alpha,i} - v_i)(v_{\alpha,j} - v_j) \\
T &= \frac{1}{n}\sum_\alpha n_\alpha \{T_\alpha + \frac{m_\alpha(\vec{v} - \vec{v}_\alpha)^2}{3k_B}\} \\
p + \frac{1}{3}\rho v^2 &= \sum_\alpha p_\alpha + \frac{1}{3}\sum_\alpha \rho_\alpha v_\alpha^2 \\
\rho\mathcal{U} + \frac{\rho v^2}{2} &= \sum_\alpha (\rho_\alpha \mathcal{U}_\alpha + \frac{\rho_\alpha v_\alpha^2}{2}) \\
P_{ij} + \rho v_i v_j &= \sum_\alpha (P_{\alpha,ij} + \rho_\alpha v_{\alpha,i} v_{\alpha,j}) \\
\Phi_k + (\rho\mathcal{U} + \frac{\rho v^2}{2})v_k + \sum_{l=1}^{3} P_{lk} v_l &= \sum_\alpha \{\Phi_{\alpha,k} + (\rho_\alpha \mathcal{U}_\alpha + \frac{\rho_\alpha v_\alpha^2}{2})v_{\alpha,k} + \sum_{l=1}^{3} P_{\alpha,lk} v_{\alpha,l}\}
\end{aligned} \tag{3.43}$$

(B1986)

14. Go back to the equation of conservation of momentum in Magnetohydrodynamics (3.31). Use three of the Maxwell's equations (2.7) and the vector identity

$$(\nabla \times \vec{u}) \times \vec{u} = (\vec{u}.\nabla)\vec{u} - \nabla(\frac{u^2}{2})$$

to write the electric force $Q\vec{E}$ and the Lorentz force $\vec{j}\times\vec{B}$ as

$$\begin{aligned} Q\vec{E} &= \epsilon_0\nabla.(\vec{E}\vec{E}) - \epsilon_0\vec{E}\times\frac{\partial\vec{B}}{\partial t} - \nabla(\frac{\epsilon_0 E^2}{2}) \\ \vec{j}\times\vec{B} &= \frac{1}{\mu_0}\nabla.(\vec{B}\vec{B}) - \nabla(\frac{B^2}{2\mu_0}) - \epsilon_0\frac{\partial\vec{E}}{\partial t}\times\vec{B} \end{aligned}$$

Combine these two results to write the electromagnetic force as

$$Q\vec{E}+\vec{j}\times\vec{B} = \nabla.\left\{-\left(\frac{\epsilon_0 E^2}{2}+\frac{B^2}{2\mu_0}\right)\mathbf{I}+\epsilon_0\vec{E}\vec{E}+\frac{\vec{B}\vec{B}}{\mu_0}\right\} - \frac{\partial}{\partial t}\left(\epsilon_0\vec{E}\times\vec{B}\right)$$

15. Show that the work per unit time (or power) done by the electromagnetic force is $\vec{j}.\vec{E}$. Start from

$$\text{Work per unit time } = \sum_\alpha \int q_\alpha(\vec{E}+\vec{w}\times\vec{B}).\vec{w} f_\alpha d^3\vec{w}$$

16. Use two of the Maxwell's equations (2.7) and the vector identity

$$\nabla.(\vec{a}\times\vec{b}) = \vec{b}.(\nabla\times\vec{a}) - \vec{a}.(\nabla\times\vec{b})$$

to find Poynting's theorem:

$$\vec{j}.\vec{E} = -\nabla.(\frac{\vec{E}\times\vec{B}}{\mu_0}) - \frac{\partial}{\partial t}(\frac{\epsilon_0 E^2}{2}+\frac{B^2}{2\mu_0})$$

17. Show that the work per unit time done by the gravitational force can be written as

$$-\rho\vec{v}.\vec{g} = \sum_{k=1}^{3}\frac{\partial(\rho\Psi v_k)}{\partial x_k} + \frac{\partial}{\partial t}(\rho\Psi) - \rho\frac{\partial\Psi}{\partial t}$$

18. Show that

$$-\vec{E}.(\vec{j}-Q\vec{v}) + (\vec{j}\times\vec{B}).\vec{v} = -\vec{E}^\star.\vec{j}^\star$$

with

$$\vec{E}^\star = \vec{E}+\vec{v}\times\vec{B},\ \ \vec{j}^\star = \vec{j}-Q\vec{v}$$

Use

$$(\vec{j}\times\vec{B}).\vec{v} = -(\vec{v}\times\vec{B}).\vec{j},\ \ (\vec{v}\times\vec{B}).\vec{v} = 0$$

James Clerk Maxwell (1831 - 1879).
One of Maxwell's most important achievements was his extension and mathematical formulation of Michael Faraday's theories of electricity and magnetic lines of force. His paper On Faraday's lines of force was read to the Cambridge Philosophical Society in two parts, 1855 and 1856. Maxwell showed that a few relatively simple mathematical equations could express the behaviour of electric and magnetic fields and their interrelation.
When the subject announced by St John's College Cambridge for the Adams Prize of 1857 was The Motion of Saturn's Rings Maxwell immediately interested. He showed that stability could be achieved only if the rings consisted of numerous small solid particles, an explanation now confirmed by the Voyager spacecraft. Maxwell's essay won him the Adams Prize and Airy wrote:- It is one of the most remarkable applications of mathematics to physics that I have ever seen.
In 1860 Maxwell was appointed to the vacant chair of Natural Philosophy at King's College in London. The six years that Maxwell spent in this post were the years when he did his most important experimental work. In London, around 1862, Maxwell calculated that the speed of propagation of an electromagnetic field is approximately that of the speed of light. He proposed that the phenomenon of light is therefore an electromagnetic phenomenon. Maxwell wrote the truly remarkable words:- We can scarcely avoid the conclusion that light consists in the transverse undulations of the same medium which is the cause of electric and magnetic phenomena.
Maxwell also continued work he had begun at Aberdeen, considering the kinetic theory of gases. By treating gases statistically in 1866 he formulated, independently of Ludwig Boltzmann, the Maxwell-Boltzmann kinetic theory of gases. This theory showed that temperatures and heat involved only molecular movement. This theory meant a change from a concept of certainty, heat viewed as flowing from hot to cold, to one of statistics, molecules at high temperature have only a high probability of moving toward those at low temperature. Maxwell's approach did not reject the earlier studies of thermodynamics but used a better theory of the basis to explain the observations and experiments.
Maxwell left King's College, London in the spring of 1865 and returned to his Scottish estate Glenlair. He made periodic trips to Cambridge and, rather reluctantly, accepted an offer from Cambridge to be the first Cavendish Professor of Physics in 1871. He designed the Cavendish laboratory and helped set it up. The Laboratory was formally opened on 16 June 1874. The four partial differential equations, now known as Maxwell's equations, first appeared in fully developed form in Electricity and Magnetism (1873). Most of this work was done by Maxwell at Glenlair during the period between holding his London post and his taking up the Cavendish chair. They are one of the great achievements of 19th-century mathematics.
On 8 October 1879 he returned with his wife to Cambridge but, by this time he could scarcely walk. One of the greatest scientists the world has known passed away on 5 November.
http://www-history.mcs.st-andrews.ac.uk/history/Mathematicians/
Article by: J. J. O'Connor and E. F. Robertson

Chapter 4

Magnetohydrodynamics

> The Sixties are often regarded as a storm that came and passed, a cyclone that blew through, its damage long repaired. But among the era's more enduring legacies was establishing a style of youth, of being young, that's been passed on for thirty years now by example in an endless chain of kids.
> *Michael Frain*
> "The survivor's Guide", May 16, 1990.
> *The Laws of our Fathers*
> Scott Turow

The attentive reader is well aware of the fact that the single-fluid equations for mass, momentum and internal energy that we obtained in the previous Chapter do not form a closed set. The aim of this Chapter is to obtain closure. Let us recall that expressions for the anisotropic part of the stress tensor and for the flux of heat due to random motions are necessary but not sufficient for obtaining this goal. Inspection of the single-fluid equations for a very simple case (3.42) revealed that there are more unknowns than equations. It dawned on us that we do not have any evolution equation for the electric current density, $\vec{j}$. Hence, we need an equation linking $\vec{j}$ to the plasma variables $\rho, P_{ij}, \vec{v}$. Since (2.6)

$$\vec{j} = \sum_{\alpha} Q_\alpha \vec{v}_\alpha = \sum_{\alpha} q_\alpha \int f_\alpha \vec{w} d\vec{w}$$

is the first order electric charge moment of f_α summed over all species α, our required equation is the first order electric charge moment of the Boltzmann equation (2.9) summed over all species α. This equation is known as the *generalized Ohm's law.* It links $\partial\vec{j}/\partial t, \vec{j}, \vec{E}, \vec{B}$ to the plasma variables. In its MHD approximation, the generalized Ohm's law links $\vec{j}, \vec{E}, \vec{B}$ to $\vec{v}$. The generalized Ohm's law contains the loss of momentum due to collisions and we cannot avoid deriving an expression for this quantity.

The generalized Ohm's law, the single-fluid equations for mass, momentum and internal energy and the pre-Maxwell equations provide us with a closed set of MHD equations. The assumptions that we need to introduce in order to derive expressions for the anisotropic part of the stress tensor, the flux of heat due to random motions and the loss of momentum due to collisions and for obtaining the approximate MHD version of the generalized Ohm's law put additional restrictions on the domain of validity of the MHD equations.

4.1 Generalized Ohm's law

In order to obtain the first charge moment of the Boltzmann equation we multiply the equation for conservation of momentum for particles of species α (3.7):

$$\begin{aligned}\frac{\partial}{\partial t}(\rho_\alpha v_{\alpha,k}) &+ \sum_{l=1}^{3}\frac{\partial}{\partial x_l}(\rho_\alpha v_{\alpha,k}v_{\alpha,l} + p_\alpha\delta_{kl} + \Pi_{\alpha,kl}) \\ &- \rho_\alpha g_k - Q_\alpha E_k - \left(\vec{j}_\alpha \times \vec{B}\right)_k = \mu_{\alpha,k}\end{aligned}$$

with q_α/m_α and sum over all species α. The result is

$$\begin{aligned}&\frac{\partial j_k}{\partial t} + \sum_\alpha\{Q_\alpha\sum_{l=1}^{3}\frac{\partial}{\partial x_l}(v_{\alpha,k}v_{\alpha,l})\} + \sum_\alpha\{\frac{q_\alpha}{m_\alpha}\sum_{l=1}^{3}\frac{\partial P_{\alpha,kl}}{\partial x_l}\} \\ &- \underbrace{Q}_{\approx 0}\,g_k - \sum_\alpha\frac{n_\alpha q_\alpha^2}{m_\alpha}(\vec{E} + \vec{v}_\alpha \times \vec{B})_k = \sum_\alpha \mu_{\alpha,k}\frac{q_\alpha}{m_\alpha}\end{aligned}$$

This is the most general form of the generalized Ohm's law. It is an exact consequence of the Boltzmann equation but not of much practical use. It is an evolution equation for the electric current density $\vec{j}$.

Let us now specialize to a fully ionized plasma consisting of electrons and ions. The index α that indicates the type of species is $\alpha = e$ for the electrons and $\alpha = i$ for the ions. In additions we assume that the ions are protons, so that $Z = 1$. The electrons have electric charge $q_e = -e$, mass m_e, their number density is n_e and their bulk velocity is $\vec{v}_e$; the ions have electric charge $q_i = e$, mass m_i, their number density is n_i and their bulk velocity is $\vec{v}_i$. In terms of these quantities we can write the macroscopic quantities $\rho, \vec{v}, Q, \vec{j}$ (3.24) as

$$\begin{aligned}\rho &= n_i m_i + n_e m_e \\ \rho\vec{v} &= n_i m_i \vec{v}_i + n_e m_e \vec{v}_e \\ Q &= (n_i - n_e)e \\ \vec{j} &= (n_i\vec{v}_i - n_e\vec{v}_e)e\end{aligned}$$

For later use we solve these equations for $n_i, n_e, \vec{v}_i, \vec{v}_e$ in terms of the global plasma quantities ρ, $\vec{v}$ and $\vec{j}$ and find

$$n_i = \frac{\rho + m_e Q/e}{m_i + m_e} \approx \frac{\rho}{m_i + m_e} \approx \frac{\rho}{m_i}$$

$$
\begin{aligned}
n_e &= \frac{\rho - m_i Q/e}{m_i + m_e} \approx \frac{\rho}{m_i + m_e} \approx \frac{\rho}{m_i} \\
\vec{v}_i &= \frac{\rho \vec{v} + m_e \vec{j}/e}{\rho + m_e Q/e} \approx \vec{v} + \frac{m_e}{\rho e}\vec{j} \approx \vec{v} + \frac{m_e}{m_i}\frac{\vec{j}}{e n_e} \approx \vec{v} \\
\vec{v}_e &= \frac{\rho \vec{v} - m_i \vec{j}/e}{\rho - m_i Q/e} \approx \vec{v} - \frac{m_i}{\rho e}\vec{j} \approx \vec{v} - \frac{\vec{j}}{e n_e}
\end{aligned} \tag{4.1}
$$

The successive approximations are obtained by first using the assumption of quasi-neutrality of the plasma and then by noting that $m_e << m_i$ and by putting consistently m_e/m_i equal to 0 compared to 1. Quasi-neutrality is a very accurate description of reality as explained in Chapter II. Quasi-neutrality means that

$$
Q \approx 0, \quad n_e \approx n_i, \quad \vec{j} = -n_e e(\vec{v}_e - \vec{v}_i)
$$

The last equality implies that the electric current is due to the relative motion of the electron gas to the ion gas. Neglecting m_e/m_i compared to 1 is not really an approximation.

Let us now rewrite the general form of the generalized Ohm's law for a fully ionized plasma consisting of electrons and protons. The loss (or gain) of momentum of the electrons due to collisions with the ions is equal to the gain (or loss) of momentum of the ions due to the same collisions. Hence,

$$
\vec{\mu}_e = -\vec{\mu}_i = \vec{R}
$$

where $\vec{R}$ is just an abbreviation. Without any approximations we get the following result

$$
\frac{m_e}{e}\frac{\partial j_k}{\partial t} + n_e m_e \sum_{l=1}^{3} \frac{\partial}{\partial x_l}(\underbrace{\frac{n_i}{n_e}}_{\approx 1} v_{i,k}v_{i,l} - v_{e,k}v_{e,l}) - \sum_{l=1}^{3}\frac{\partial}{\partial x_l}(P_{e,kl} - \underbrace{\frac{m_e}{m_i}}_{<<1} P_{i,kl})
$$

$$
-en_e(1 + \underbrace{\frac{n_i}{n_e}}_{\approx 1}\underbrace{\frac{m_e}{m_i}}_{<<1})E_k - en_e\{(\vec{v}_e + \underbrace{\frac{n_i}{n_e}}_{\approx 1}\underbrace{\frac{m_e}{m_i}}_{<<1}\vec{v}_i) \times \vec{B}\}_k = -R_k(1 + \underbrace{\frac{m_e}{m_i}}_{<<1})
$$

This equation can be simplified by using the approximations $m_e/m_i = 0$, $n_i/n_e = 1$. The quadratic term in the velocities $(v_{ik}v_{il} - v_{ek}v_{el})$ can be rewritten in terms of the global plasma quantities $\vec{v}, \vec{j}$ by using the expressions for $\vec{v}_i, \vec{v}_e$ in terms of $\vec{v}, \vec{j}$. The result is

$$
\frac{m_e}{e^2 n_e}\left\{\frac{\partial j_k}{\partial t} + \sum_{l=1}^{3}\frac{\partial}{\partial x_l}(v_k j_l + v_l j_k - \frac{1}{en_e}j_k j_l)\right\} - \frac{1}{en_e}\sum_{l=1}^{3}\frac{\partial P_{e,kl}}{\partial x_l} + \frac{R_k}{en_e} = (\vec{E} + \vec{v}_e \times \vec{B})_k
$$

Let us note again that this is an evolution equation for the electric current $\vec{j}$. It contains the terms $P_{e,kl}$, $\mu_{e,i}$ for which we do not have expressions so far.

OK, let us now get down to the real approximations. So far we have only assumed that the plasma is electrically quasi-neutral and that the mass of the electron can be ignored compared to that of an ion. These are not really approximations to be worried about. In order to make further progress in Ohm's law we need to have an expression for the transfer of momentum from the electrons due to the collisions with the ions. Another name for this transfer of momentum is the drag force on the electron fluid due to the electron collisions with the ions. This drag force was already introduced in (2.55) where it was assumed to be proportional to the to the relative velocity of the electron gas to the ion gas. Hence, we can borrow (2.55) and note that here $\vec{\mu}_e = \vec{F}_{fr}$, so that

$$\vec{\mu}_e = -\vec{\mu}_i = \vec{R} = -\overline{\nu}_{e,i} n_e m_e (\vec{v}_e - \vec{v}_i)$$
$$= \overline{\nu}_{e,i} \frac{m_e}{e} \vec{j} = e n_e \frac{\vec{j}}{\sigma} = e n_e \tilde{\eta} \vec{j} \qquad (4.2)$$

$\overline{\nu}_{e,i}$ is the average electron collision frequency defined in Chapter II. According to (2.56) it varies inversely with $T_e^{3/2}$ and is independent of the ion mass. $\tilde{\eta}$ is the electrical resistivity and σ is the electrical conductivity. They are defined as (2.57)

$$\tilde{\eta} = \frac{m_e \overline{\nu}_{e,i}}{n_e e^2}, \quad \sigma = \frac{1}{\tilde{\eta}} \qquad (4.3)$$

We can now write the generalized Ohm's law as

$$E_k = -(\vec{v}_e \times \vec{B})_k + \frac{j_k}{\sigma} - \frac{1}{e n_e} \sum_{l=1}^{3} \frac{\partial P_{e,kl}}{\partial x_l} + \frac{m_e}{e^2 n_e} \left\{ \frac{\partial j_k}{\partial t} + \sum_{l=1}^{3} \frac{\partial}{\partial x_l} (v_k j_l + v_l j_k - \frac{1}{e n_e} j_k j_l) \right\}$$

Dyadic notation helps us to write the generalized Ohm's law in a more compact form as

$$\vec{E} = -(\vec{v}_e \times \vec{B}) + \frac{\vec{j}}{\sigma} - \frac{1}{e n_e} \nabla . \mathbf{P_e} + \frac{m_e}{e^2 n_e} \left\{ \frac{\partial \vec{j}}{\partial t} + \nabla . (\vec{\mathbf{v}}\vec{\mathbf{j}} + \vec{\mathbf{j}}\vec{\mathbf{v}} - \frac{1}{e n_e} \vec{\mathbf{j}}\vec{\mathbf{j}}) \right\} \qquad (4.4)$$

We promised that we would derive an equation that links links $\partial \vec{j}/\partial t, \vec{j}, \vec{E}, \vec{B}$ to the plasma variables. This requires elimination of $\vec{v}_e$ and of $\mathbf{P_e}$. This is easily done for $\vec{v}_e$ and introduces the Hall term as an additional term in the right hand side of (4.4)

$$\vec{E} = -(\vec{v} \times \vec{B}) + \frac{\vec{j}}{\sigma} + \frac{1}{e n_e} \vec{j} \times \vec{B} - \frac{1}{e n_e} \nabla . \mathbf{P_e}$$
$$+ \frac{m_e}{e^2 n_e} \left\{ \frac{\partial \vec{j}}{\partial t} + \nabla . (\vec{\mathbf{v}}\vec{\mathbf{j}} + \vec{\mathbf{j}}\vec{\mathbf{v}} - \frac{1}{e n_e} \vec{\mathbf{j}}\vec{\mathbf{j}}) \right\} \qquad (4.5)$$

The generalized Ohm's is written as an equation for the electric field and not as an equation for $\partial \vec{j}/\partial t$. The reason is that the contribution containing the latter term will be dropped in a few minutes. The terms on the right-hand side all have specific names:

- $$-(\vec{v} \times \vec{B}) = \text{ the dynamo term or the induction term}$$

It represents the electric field generated by the motion of the plasma as a whole in the presence of a magnetic field.

- $$\frac{\vec{j}}{\sigma} = \text{ Ohmic term}$$

- $$\frac{1}{e n_e} \vec{j} \times \vec{B} = \text{ the Hall term}$$

- $$-\frac{1}{en_e}\nabla.\mathbf{P_e} = \text{ the battery term}$$

It is the electric field that tries to counteract the motions of the electrons driven by the press tensor in order to keep the plasma electrically neutral.

- $$\frac{m_e}{e^2 n_e}\left\{\frac{\partial \vec{j}}{\partial t} + \nabla.(\vec{\mathbf{v}}\vec{\mathbf{j}} + \vec{\mathbf{j}}\vec{\mathbf{v}} - \frac{1}{en_e}\vec{\mathbf{j}}\vec{\mathbf{j}})\right\} = \text{ the electron inertia term.}$$

In view of the very small mass of the electrons we can drop this term without any problem.

An important assumption in classic Magnetohydrodynamics is that the collisions lead to isotropic distribution functions for both electrons and ions. Hence the pressure tensors for electrons and ions (3.25) have only contributions from the isotropic electron pressure and the isotropic ion pressure:

$$P_{e,kl} = p_e\delta_{kl}, \quad P_{i,kl} = p_i\delta_{kl}, \quad P_{kl} = p\delta_{kl}, \quad \Pi_{kl} = 0 \qquad (4.6)$$

From a mathematical point of view the quantities T_e, T_i and T are allowed to differ. However from a physical point of view the situation is different. If a single-fluid theory is to make physical sense, the ion and electrons fluids should have comparable temperatures. Ideally, the ions and electrons should have the same temperature which is the temperature of the plasma as a whole T

$$T_e = T_i = T \qquad (4.7)$$

We shall come back on the assumptions (4.6) and (4.7) in Section 4.4. when we scrutinize the conditions for validity of MHD. Combined with $n_e = n_i$ and $n = n_e + n_i$, (4.7) implies that

$$p = nk_BT, \quad p_e = n_ek_BT_e = n_ik_BT_i = p_i, \quad p = p_e + p_i \qquad (4.8)$$

The reader is now asked to go back to Problem 13 of the previous Chapter. Reduce the second line of this problem for a two-fluid plasma to

$$p = p_e + p_i + \frac{\rho_e}{3} \mid \vec{v}_e - \vec{v} \mid^2 + \frac{\rho_i}{3} \mid \vec{v}_i - \vec{v} \mid^2$$

Since

$$\vec{v}_i \approx \vec{v} + \frac{m_e}{m_i}\frac{\vec{j}}{en_e} \approx \vec{v}, \quad \vec{v}_e \approx \vec{v} - \frac{\vec{j}}{en_e}$$

writing

$$p = p_e + p_i$$

comes down on assuming that

$$\mid \vec{v}_e - \vec{v}_i \mid << \mid \vec{v}_i \mid$$

so that the electron and ion fluids move at nearly equal velocities. This condition will come back when discussing the MHD approximation of the generalized Ohm's law. With the pressure tensor reduced to an isotropic pressure we can write the battery term as

$$-\frac{1}{en_e}\nabla p_e.$$

When we drop the electron inertia term, we can write the generalized Ohm's law as

$$\vec{E} = -(\vec{v}\times\vec{B}) + \frac{\vec{j}}{\sigma} + \frac{1}{en_e}(\vec{j}\times\vec{B}) - \frac{1}{en_e}\nabla p_e \qquad (4.9)$$

When we compare the generalized Ohm's law with the well-known Ohm's law for an electric conductor in rest

$$\vec{j} = \sigma\vec{E}, \quad \vec{E} = \frac{\vec{j}}{\sigma}$$

we immediately understand the meaning of the adjective "generalized".

4.2 The MHD approximation of Ohm's law

The generalized Ohm's law (4.9) is written as an equation for the electric field and has four terms on its right-hand side. Rather complicated! Let us find out which terms are important and which ones can be discarded. In order to do so we choose the induction term $\vec{v}\times\vec{B}$ as reference and compare the other terms to this reference term. This comparison is done by scale analysis. We keep in mind that MHD is a mathematical theory of non-relativistic macroscopic plasma phenomena. Macroscopic means that the typical length scale L of the spatial variations and the typical time scale τ of the temporal variations that we are interested in, are orders of magnitudes longer than the microscopic length scales and time scales defined by the electron and ion dynamics. Often we use $\omega = 1/\tau$, which has the dimensions of as frequency, instead of τ for characterizing the large-scale low-frequency fluid-like behaviour of a plasma system. In Chapter II we found that the quantities $\omega_{p,e}, \lambda_D, \omega_{c,e}, r_{L,e}$ characterize the high-frequency and short scale dynamics of the electrons and that the quantities $\omega_{p,i}, \omega_{c,i}, r_{L,i}$ are characteristic for the the low-frequency dynamics of the ions. Non-relativistic means that we are dealing with velocities $v \approx L/\tau << c$. The scale analysis is based on using L and τ for obtaining (crude) approximations of temporal and spatial derivatives. Formally we replace derivatives using the simple rule

$$\frac{\partial}{\partial t} \longrightarrow \frac{1}{\tau}, \quad \nabla \longrightarrow \frac{1}{L}$$

For example

$$\mid \nabla\times\vec{B} \mid\approx \frac{B}{L}, \quad \mid \frac{\partial\vec{E}}{\partial t} \mid\approx \frac{E}{\tau}$$

Let us first go back to the third of Maxwell's equations (2.7). It is straightforward to show that the displacement current can be neglected compared to $\nabla\times\vec{B}$ in a non-relativistic theory. We use the first of Maxwell's equations (2.7) to estimate E/B as

$$\frac{E}{B} \approx \frac{L}{\tau}$$

Hence

$$\frac{|(1/c^2)\partial\vec{E}/\partial t|}{|\nabla\times\vec{B}|} \approx \frac{1}{c^2}\frac{E}{B}\frac{L}{\tau} \approx \frac{1}{c^2}\left(\frac{L}{\tau}\right)^2 \approx (\frac{v}{c})^2 << 1$$

In a non-relativistic theory all terms proportional to $(v/c)^2$ are neglected and consequently the displacement current can be neglected and the third of Maxwell's equations (2.7) takes the simple form

$$\vec{j} = \frac{1}{\mu}\nabla\times\vec{B} \qquad (4.10)$$

Let us now compare the Ohmic term to the induction term as

$$\frac{|\text{ Ohmic term }|}{|\text{ Induction term }|} = \frac{|\tilde{\eta}\vec{j}|}{|\vec{v}\times\vec{B}|} = \frac{\tilde{\eta}}{\mu}\frac{|\nabla\times\vec{B}|}{|\vec{v}\times\vec{B}|} \approx \frac{\tilde{\eta}}{\mu L v} = \frac{1}{\sigma\mu L v} = \frac{1}{R_m}$$

R_m is the *magnetic Reynolds number* and is defined as

$$R_m = \sigma\mu L v = \frac{Lv}{\eta} \qquad (4.11)$$

where η is defined as

$$\eta = \frac{1}{\mu\sigma} = \frac{\tilde{\eta}}{\mu} \qquad (4.12)$$

It is named the magnetic diffusivity and should not be confused with $\tilde{\eta}$; $\eta \neq \tilde{\eta}$.

The magnetic Reynolds number (4.11) is almost always a very large number. It is the product of the electrical conductivity σ, the length scale L and the velocity v. An increase in any of these three quantities leads to larger values for R_m. Plasma flow velocities are usually restricted to low values. In tokamak plasma physics R_m is large because the electrical conductivity is large. In solar physics and astrophysics R_m is large, even for relatively moderate electrical conductivities, because of the large spatial dimensions of the systems. In that sense we are dealing with a *large-scale limit.*

In what follows we shall have to make choices for the velocity v which is to be understood as a velocity characterizing the very-low-frequency fluid-like behaviour of plasmas. In Chapter II we have seen that the quantities $\omega_{p,e}, \lambda_D, \omega_{c,e}, r_{L,e}$ characterize the high-frequency and very short scale behaviour of a plasma dictated by the dynamics of the electrons and that the quantities $\omega_{p,i}, \omega_{c,i}, r_{L,i}$ are characteristic for the low-frequency behaviour of a plasma dictated by the dynamics of the ions. In Section 8 of the present Chapter and throughout Chapter V we shall see that the very-low-frequency fluid-like dynamics of the magnetic field lines to which the plasma is fastened is characterized by the Alfvén velocity v_A. It is defined as

$$v_A^2 = \frac{B^2}{\mu\rho} \qquad (4.13)$$

In Chapter V the sound velocity, v_S, defined as

$$v_S^2 = \frac{\gamma p}{\rho} \approx \frac{2\gamma p_i}{\rho_i} = 2\gamma v_{t,i}^2 \qquad (4.14)$$

will emerge as the velocity that characterizes the dynamics associated with the plasma pressure and compressibility. The ratio of the plasma pressure p to the magnetic pressure $B^2/2\mu$ is known as the plasma beta

$$\beta = \frac{p}{B^2/2\mu} \qquad (4.15)$$

The very-low-frequency fluid-like behaviour of a plasma is characterized by the quantities v_A, v_S, L, where L is a length scale for the system as a whole which can be specified by the spatial variations of the global quantities.

Let us now go back to the magnetic Reynolds number R_m (4.11). If we substitute the Alfvén velocity v_A for the velocity v in the definition of R_m then R_m is called the Lundquist number Lu

$$Lu = \sigma\mu L v_A = \frac{L v_A}{\eta} \qquad (4.16)$$

Let us now compare the Hall term to the induction term as

$$\frac{|\text{ Hall term }|}{|\text{ Induction term }|} = \frac{1}{en_e}\frac{|\vec{j}\times\vec{B}|}{|\vec{v}\times\vec{B}|} \approx \frac{1}{en_e}\frac{|\vec{j}|}{|\vec{v}|} \approx \frac{1}{en_e}\frac{B}{\mu v L}$$

Recall that $n_e = n_i = \frac{\rho}{m_i}$ so that

$$\frac{1}{en_e} = \frac{B/\rho}{\omega_{c,i}}$$

where $\omega_{c,i} = eB/m_i$ is the Larmor frequency or cyclotron gyration frequency for ions (2.65). Hence

$$\frac{|\text{ Hall term }|}{|\text{ Induction term }|} \approx \frac{v/L}{\omega_{c,i}}\frac{B^2/\mu}{\rho v^2}$$

In order to further evaluate this ratio we need to make a choice for v. For a plasma with its internal energy of the same order as the magnetic energy the thermal velocity of the ions $v_{t,i}$ is a good choice for the characteristic velocity v. Hence,

$$\frac{|\text{ Hall term }|}{|\text{ Induction term }|} \approx \frac{v_{t,i}/L}{\omega_{c,i}}\underbrace{\frac{B^2/\mu}{\rho v_{t,i}^2}}_{\approx 1} \approx \frac{r_{L,t,i}}{L}$$

We can drop the Hall term in comparison to the induction term if

$$\frac{r_{L,t,i}}{L} << 1 \qquad (4.17)$$

This is exactly condition (2.70) for the ions to be magnetized. In Chapter II we have found the equivalent conditions (2.73). The Hall term can be dropped in comparison to the induction term if the ions are magnetized. The ions have to be effectively glued to the magnetic field lines and should consequently perform many gyro-rotations during the ion thermal transit time. This is a strong condition as was explained in Chapter II. Recall also that magnetized ions imply that the electrons are also magnetized, so that in order for the Hall term to be dropped the plasma as a whole must be glued to the magnetic field lines.

For a plasma with its internal energy much smaller than the magnetic energy, the Alfvén velocity is a good choice for the characteristic velocity v. Hence,

$$\frac{\mid \text{Hall term} \mid}{\mid \text{Induction term} \mid} \approx \frac{v_A/L}{\omega_{c,i}} \underbrace{\frac{B^2/\mu}{\rho v_A^2}}_{=1} \approx \frac{1}{\tau_A \omega_{c,i}} \approx \frac{r_{L,A,i}}{L}$$

with τ_A the Alfvén transit time and $r_{L,A,i}$ the Larmor gyration radius for ions moving at the Alfvén velocity

$$\tau_A = \frac{L}{v_A}, \quad r_{L,A,i} = \frac{v_A}{\omega_{c,i}}$$

The conclusion is the same. The Hall term can be dropped in comparison to the induction term if

$$\boxed{\frac{r_{L,A,i}}{L} << 1} \qquad (4.18)$$

The Larmor gyration radius of the ions has to be much smaller than the length scale of the system so that the ions are effectively fastened to the magnetic field lines. An equivalent condition is that the ions perform many gyro-rotations during the Alfvén transit time. The ions have to be magnetized! This implies that the electrons are also magnetized. Hence, we can drop the Hall term if the plasma as a whole is glued to the magnetic field lines.

Another estimate for the relative importance of the Hall term compared to the induction term follows from

$$\vec{j} = -n_e e(\vec{v}_e - \vec{v}_i) \quad \Rightarrow \frac{\vec{j}}{en_e} = \vec{v}_i - \vec{v}_e = \vec{v} - \vec{v}_e$$

so that

$$\frac{\mid \text{Hall term} \mid}{\mid \text{Induction term} \mid} \approx \frac{\mid \vec{v} - \vec{v}_e \mid}{\mid \vec{v} \mid}$$

The Hall term can be dropped in comparison to the induction term if

$$\boxed{\frac{\mid \vec{v} - \vec{v}_e \mid}{\mid \vec{v} \mid} << 1} \qquad (4.19)$$

Hence we can drop the Hall term if the drift velocity of the electron gas with respect to the plasma as a whole is small. The electrons are easily magnetized and consequently tightly glued to the magnetic field lines. Saying that the drift velocity of the electron gas to the plasma as a whole is small, implies that the ions also should be fastened to the magnetic field lines. Hence, the Hall term can be dropped if the plasma as a whole is fastened to the magnetic field lines. Go back to (4.8) and look at the condition for the approximation $p = p_e + p_i$ to be valid. Remarkable, isn't it?

Compare now the battery term to the induction term as

$$\frac{|\text{ Battery term }|}{|\text{ Induction term }|} = \frac{|\nabla p_e/en_e|}{|\vec{v}\times\vec{B}|} \approx \frac{1}{en_e}\frac{p_e}{L}\frac{1}{vB}$$

We use $p_e = n_e k_B T_e$, $n_e = n_i$ and note that MHD implicitly assumes that $T_e = T_i$ so that $p_e = p_i = n_i m_i v_{t,i}^2$. Hence

$$\frac{|\text{ Battery term }|}{|\text{ Induction term }|} \approx \frac{m_i}{eB}\frac{1}{L}\frac{v_{t,i}^2}{v} \approx \frac{v_{t,i}}{\omega_{c,i}}\frac{1}{L}\frac{v_{t,i}}{v} \approx \frac{r_{L,t,i}}{L}\frac{v_{t,i}}{v}$$

O.K. what about v? As for the Hall term we choose the thermal velocity of the ions $v_{t,i}$ if the internal energy of the plasma is of the same order as the magnetic energy. In that case

$$\frac{|\text{ Battery term }|}{|\text{ Induction term }|} \approx \frac{r_{L,t,i}}{L}$$

The condition for the electron pressure term to be dropped is then the same as that for the Hall term to be dropped (4.17); the ions have to be fastened to the magnetic field lines. In case the internal energy of the plasma is much smaller than the magnetic energy we use v_A as characteristic velocity. Hence,

$$\frac{|\text{ Battery term }|}{|\text{ Induction term }|} << \frac{r_{L,t,i}}{L}$$

Hence for strong magnetic fields the condition for the electron pressure term to be dropped is much less stringent.

Let us now compare the electron inertia term to the induction term. We can drop this term based on the simple argument that the electron mass m_e is very small and that the inertial effects on the electrons can be neglected all together. A more educated argument goes as follows. We compare the first contribution to the electron inertia term with the left hand side of the generalized Ohm's law $\vec{E}$

$$\frac{|\left(m_e/e^2 n_e\right)\partial\vec{j}/\partial t|}{|\vec{E}|} \approx \frac{m_e}{e^2 n_e}\frac{1}{\tau}\frac{|\vec{j}|}{|\vec{E}|}$$

We use the fourth of Maxwell's equations (2.7) and the equation for conservation of charge

$$|E| \approx L\frac{|Q|}{\epsilon}, \quad |\vec{j}| \approx \frac{L}{\tau}|Q|$$

and obtain

$$\frac{|\left(m_e/e^2 n_e\right)\partial\vec{j}/\partial t|}{|\vec{E}|} \approx \frac{\epsilon m_e}{e^2 n_e}\frac{1}{\tau^2} \approx \frac{1}{\omega_{p,e}^2\tau^2} \approx \frac{\omega^2}{\omega_{p,e}^2} << 1$$

$\omega = 1/\tau$ is a macroscopic frequency associated with the slow variation on the long time scale τ. For plasma phenomena with a time scale that is sufficiently longer than the plasma oscillation period, that is for all macroscopic (low-frequency) plasma phenomena, the electron inertia term can be dropped.

Let us come back to the Hall term and the battery term or electron pressure term. An estimate for the ratio of these two terms is as follows

$$\frac{|\text{ Battery term }|}{|\text{ Hall term }|} = \frac{|\nabla p_e/en_e|}{|\vec{j}\times\vec{B}/en_e|} = \frac{|\nabla p_e|}{|\vec{j}\times\vec{B}|} \approx \frac{p_e}{B^2/\mu}\frac{L_B}{L_p} \approx \beta\frac{L_B}{L_p}$$

L_B and L_p are the length scales for the variation of $\vec{B}$ and p_e respectively. Usually their ratio is of the order of unity. The relative importance of the battery term to the Hall term is controlled by the plasma β. For a sufficiently strong magnetic field the battery term is unimportant compared to the Hall term.

There is no such thing as a set of typical values for v, L, n_e, B, T_e for space plasmas. You can look at plasmas in the solar chromosphere, the solar corona, the solar wind, the magnetosphere of the earth and on each occasion you are confronted with a different set. We take

$$v = 10^5\text{m/s},\ \ L = 10^6\text{m},\ \ n_e = 10^7\text{m}^{-3},$$

$$B = 10^{-8}\text{Tesla},\ \ T_e = 10^5\text{K}$$

as a set of typical values (taken from S1983) and use

$$\mu_0 = 4\pi\times 10^{-7}\text{H m}^{-1},\ \ e = 1.6022\times 10^{-19}\text{C},$$

$$m_e = 9.1094\times 10^{-31}\text{kg},\ \ k_B = 1.3807\times 10^{-23}\text{J K}^{-1}$$

We then find that

$$\frac{|\text{ Ohmic term }|}{|\text{ Induction term }|} = 10^{-12} \ll 1$$

$$\frac{|\text{ Hall term }|}{|\text{ Induction term }|} \approx 0.1$$

$$\frac{|\text{ Battery term }|}{|\text{ Induction term }|} = 10^{-2} \ll 1$$

$$\frac{|\text{ Electron inertia term }|}{|\text{ Induction term }|} \approx 10^{-5} \ll 1$$

This example shows that in a typical large-scale low-frequency space plasma situation, the induction term $-\vec{v}\times\vec{B}$ is comfortably larger than all the other terms on the right hand side of the generalized Ohm's law (4.5), with the possible exception of the Hall term, $\vec{j}\times\vec{B}/(en_e)$. This implies that a very good approximation to the generalized Ohm's law (4.5) is

$$\vec{E} = -\vec{v}\times\vec{B} \qquad (4.20)$$

This is the *ideal MHD approximation or the ideal MHD limit* of the generalized Ohm's law. It is an equation for the electric field in terms of the plasma velocity $\vec{v}$ and the magnetic field $\vec{B}$ and shows that the electric field is generated by the motion of the plasma as a whole in the presence of a magnetic field. At this point the reader should be aware of the fact that the estimates used to arrive at the ideal MHD approximation of the generalized Ohm's law, are global estimates. Local deviations can, and as a matter of fact, do occur. Note that $\vec{E}^\star = \vec{E} + \vec{v} \times \vec{B}$ is the electric field seen by an observer who is moving with the plasma velocity. In ideal MHD $\vec{E}^\star = 0$.

We can obtain a more accurate approximation for the electric field by retaining the Hall term (4.5)

$$\vec{E} = -\vec{v}_e \times \vec{B} \qquad (4.21)$$

This is Ohm's law in *Hall MHD.* This more accurate version of the MHD approximation of Ohm's law shows that the electric field is caused by the motion of the electron gas in the presence of a magnetic field.

When we retain the Ohmic term $\vec{j}/\sigma$ in addition to the induction term, the generalized Ohm's law is

$$\vec{E} = -\vec{v} \times \vec{B} + \frac{\vec{j}}{\sigma}$$

which we can rewrite as

$$\vec{j} = \sigma\left(\vec{E} + \vec{v} \times \vec{B}\right) = \sigma\vec{E}^\star \qquad (4.22)$$

This is the generalized Ohm's law of *resistive MHD.* It is an algebraic relation between $\vec{j}, \vec{E}, \vec{B}, \vec{v}$. It is the generalization of the classic Ohm's law for a conductor in rest, since $\vec{E}^\star = \vec{E} + \vec{v} \times \vec{B}$ is the electric field measured in a reference frame that moves with velocity $\vec{v}$. $\vec{E}^\star$ is the electric field seen by an observer who moves with the plasma. From a formal point of view we can obtain the ideal MHD approximation of Ohm's law by taking the limit $\sigma = \infty$ in the resistive MHD approximation of Ohm's law. This mathematical fact has led people to talk about infinitely electrically conducting plasma. From a physical point of view this does not make sense. The ideal MHD approximation is based on the very large values of the magnetic Reynolds number and these large numbers occur in solar and space plasmas because of the huge spatial dimensions of these plasmas, not because of their high electrical conductivities.

The nature of Ohm's law in ideal MHD is different from that in resistive MHD. In resistive MHD, it is an equation that expresses $\vec{j}$ in terms of $\vec{E}$, $\vec{B}$ and $\vec{v}$; in ideal MHD it is an equation that expresses $\vec{E}$ in terms of $\vec{B}$ and $\vec{v}$. As a consequence we need an additional equation for computing $\vec{j}$ in ideal MHD.

Two final comments in this Section. The first comment is concerned with the battery term which we shall largely forget in what follows. Dimensional analysis has convinced us that $\mid \nabla p_e/(en_e) \mid \ll \mid \vec{v} \times \vec{B} \mid$. The induction term $-\vec{v} \times \vec{B}$ lies in a plane normal to $\vec{B}$ and has no component along $\vec{B}$. ∇p_e has components parallel and normal to $\vec{B}$. Hence, the battery term can generate a (weak) electrical field in the direction along $\vec{B}$ that we shall not be concerned about. The second comment is concerned with the resistive term in Ohm's law. Dimensional analysis indicates that it is overall very small compared to the induction term and also compared the Hall term. The reason why the resistive term is often kept in favour of the Hall term is that the resistive term causes diffusion of the magnetic field and dissipation of magnetic energy in heat.

4.3 The pre-Maxwell equations

Let us go back to the Maxwell's equations (2.7) for the electromagnetic field. In the previous Section we have seen that for non-relativistic phenomena the displacement current can be dropped from the third equation of this set producing the simple relation (4.10) between $\vec{j}$ and $\vec{B}$.

Quasi-neutrality is a key element in MHD. In Chapter II we have shown that quasi-neutrality is a very good approximation of reality on length scales and time scales that are sufficiently longer than the Debye length and the plasma electron oscillation period. Let us have a look at charge separation and quasi-neutrality again and see how they tie in with the ideal MHD approximation of Ohm's law.

We compare the total charge Q to the charge of the electrons and show that this ratio is much smaller than 1. We use the fourth of the Maxwell equations (2.7) $Q = \epsilon \nabla . \vec{E}$ to estimate Q as

$$\mid Q/(en_e) \mid = \mid \epsilon \nabla . \vec{E}/(en_e) \mid \approx \frac{\epsilon}{en_e} \frac{E}{L} \approx \frac{\epsilon}{en_e} \frac{vB}{L}$$

The non-relativistic version of the third of the Maxwell equations (4.10) can be used to estimate B/L as

$$\frac{B}{L} \approx \mu \mid \vec{j} \mid \approx \mu e n_e \mid \vec{v} - \vec{v}_e \mid$$

and find

$$\mid Q/(en_e) \mid \approx \epsilon \mu v \mid \vec{v} - \vec{v}_e \mid \approx \frac{v \mid \vec{v} - \vec{v}_e \mid}{c^2} \ll 1$$

Charge neutrality is a very good approximation of reality in non-relativistic theory. We can obtain the same conclusion by comparing the total charge Q to the charge of the ions.

$$\mid Q/(Zen_i) \mid \approx \frac{\epsilon}{Zen_i} \frac{vB}{L} \approx \frac{\epsilon}{\omega_{c,i}} \frac{B}{\rho} \frac{vB}{L} \approx \frac{\epsilon\mu}{\omega_{c,i}} \frac{B^2}{\mu\rho} \frac{v}{L} \approx \frac{\omega}{\omega_{c,i}} \frac{v_A^2}{c^2} \ll 1$$

Here $\omega = v/L$ is a macroscopic, very low frequency.

Let us now see how quasi-neutrality affects other quantities and equations in MHD. In Chapter III (3.37) we have seen that the electric current $\vec{j}$ is the sum of the convective electric current density $Q\vec{v}$, caused by the net charge density that is convected away by the flow of the plasma as a whole, and by the conductive electric current density $\vec{j}^\star$, caused by the relative motions of the positive and negative charges. We now show that the convective electric current density $Q\vec{v}$ can be neglected compared to the conductive electric current density $\vec{j}^\star$. Use $\nabla . \vec{E} = Q/\epsilon$, $\vec{E} = -\vec{v} \times \vec{B}$, $\vec{j} = (\nabla \times \vec{B})/\mu$ to find

$$\frac{\mid Q\vec{v} \mid}{\mid \vec{j} \mid} \approx \frac{\epsilon v^2 B/L}{B/(L\mu)} \approx \epsilon\mu v^2 \approx \left(\frac{v}{c}\right)^2 \ll 1$$

The simplified non-relativistic MHD versions of the equations of Maxwell or the so-called *pre-Maxwell equations* are

$$\nabla \times \vec{E} = -\frac{\partial \vec{B}}{\partial t}$$

$$\nabla . \vec{B} = 0$$

$$\nabla \times \vec{B} = \mu \vec{j}$$

$$\nabla . \vec{E} = \frac{Q}{\epsilon} \qquad (4.23)$$

We have retained the right hand side Q/ϵ in the 4th Maxwell equation in stead of putting it equal to 0. I have no doubt that the attentive reader has figured out why these equations are called the pre-Maxwell equations. Neglect of the relativistic contributions to Maxwell's equations has removed the electromagnetic waves and sent us back to the era before Maxwell.

4.4 Equations of Ideal and resistive MHD

The equation of motion

We now show that the electrostatic force $Q\vec{E}$ in the equation of motion can be neglected in comparison with the Lorentz force $\vec{j} \times \vec{B}$. We use the following intermediate results

$$|\, Q \,| \approx \epsilon E/L, \;\; E \approx vB, \;\; |\, Q\vec{E} \,| \approx \epsilon v^2 B^2/L$$

$$B/L \approx \mu \,|\, \vec{j} \,|, \;\; |\, Q\vec{E} \,| \approx \epsilon \mu v^2 B \,|\, \vec{j} \,| \approx (v/c)^2 B \,|\, \vec{j} \,|$$

$$|\, \vec{j} \times \vec{B} \,| \approx B \,|\, \vec{j} \,|, \;\; \frac{|\, Q\vec{E} \,|}{|\, \vec{j} \times \vec{B} \,|} \approx (v/c)^2 \ll 1$$

The conclusion is clear. In a consistent non-relativistic theory we have to throw the electrostatic force out of the equation of motion in comparison to Lorentz force.

A similar result can be obtained for the energy densities in the electrostatic field and in the magnetic field:

$$\frac{\epsilon E^2}{B^2/\mu} = \epsilon\mu \frac{E^2}{B^2} \approx \epsilon\mu \frac{v^2 B^2}{B^2} \approx (v/c)^2 \ll 1$$

Let us recall that classic Magnetohydrodynamics assumes that collisions lead to isotropic distribution functions for both electrons and ions. Hence the pressure tensors for electrons and ions have only contributions from the isotropic electron pressure and from the isotropic ion pressure (4.6). In addition it is implicitly accepted that $T_e = T_i = T$ (4.7) so that $p = nk_BT = p_e + p_i$ (4.8). Reducing the pressure tensor to the isotropic pressure means that viscosity is neglected. The equation of motion in non-viscous MHD is:

$$\rho \frac{d\vec{v}}{dt} = -\nabla p + \rho \vec{g} + \vec{j} \times \vec{B} \qquad (4.24)$$

Anisotropic transport coefficients

The electrical conductivity σ that appears in the generalized Ohm's law of resistive MHD (4.2) has been treated as a scalar so far. In Section 2.8 I have explained that the motions of charged particles in the presence of a magnetic field can be drastically different from those in an nonmagnetic plasma. In a nonmagnetic plasma life is the same in all directions and there is perfect isotropy. As a consequence the transport coefficients are scalars. In a strong magnetic field the charged particles are effectively glued to the magnetic field lines. Because of the influence of the magnetic field on the particle trajectories, a magnetic plasma is not isotropic, having instead a limited form of isotropy in which the transport coefficients are invariant only for rotations at a given point about a line parallel to $\vec{B}$. The transport coefficients are tensors that are laterally isotropic about $\vec{B}$. Scalar transport coefficients apply to nonmagnetic plasmas or weakly magnetized plasmas in the sense that

$$\overline{\nu}_{e,i} >> \omega_{c,e} \tag{4.25}$$

However, in plasmas with strong magnetic fields where

$$\overline{\nu}_{e,i} << \omega_{c,e} \tag{4.26}$$

the dissipation coefficients are no longer isotropic but rather anisotropic and described by tensor. It can be anticipated that the scalar dissipation coefficients obtained for nonmagnetic plasmas or weakly magnetized plasmas, describe the behaviour parallel to the magnetic field lines. We illustrate the anisotropy of the dissipation coefficients for the electrical conductivity. Our starting point is a reduced form of the equation of motion of the electron fluid in which the electron inertia, pressure and viscosity are dropped. In the plasma reference frame ($\vec{v} = \vec{v}_i$), this equation is

$$-e\vec{E}^{\star} - e\vec{v}_e \times \vec{B} - m_e \overline{\nu}_{e,i} \vec{v}_e = 0 \tag{4.27}$$

Since in the plasma reference frame $\vec{j} = -en_e\vec{v}_e$ the previous equation (4.27) can be rewritten as

$$\vec{j} = \sigma_0 \vec{E}^{\star} - \frac{\sigma_0}{en_e} \vec{j} \times \vec{B}$$

where σ_0 is the isotropic scalar conductivity defined in (4.3). In order to make further progress we take the z-axis along the magnetic field and rewrite the previous equation as

$$\begin{aligned} j_x &= \sigma_0 E_x^{\star} - \frac{\omega_{c,e}}{\overline{\nu}_{e,i}} j_y \\ j_y &= \sigma_0 E_y^{\star} + \frac{\omega_{c,e}}{\overline{\nu}_{e,i}} j_x \\ j_z &= \sigma_0 E_z^{\star} \end{aligned} \tag{4.28}$$

This set of equations (4.28) is now considered as a set of three linear equations for the unknown quantities j_x, j_y, j_z. Its solution is

$$\vec{j} = \sigma \vec{E}^{\star} \tag{4.29}$$

σ is the conductivity tensor

$$\sigma = \begin{bmatrix} \sigma_\perp & -\sigma_\wedge & 0 \\ \sigma_\wedge & \sigma_\perp & 0 \\ 0 & 0 & \sigma_\parallel \end{bmatrix} \tag{4.30}$$

where

$$\sigma_\parallel = \sigma_0, \quad \sigma_\perp = \frac{\overline{\nu}_{e,i}^2}{\omega_{c,e}^2 + \overline{\nu}_{e,i}^2}\sigma_0, \quad \sigma_\wedge = \frac{\omega_{c,e}\overline{\nu}_{e,i}}{\omega_{c,e}^2 + \overline{\nu}_{e,i}^2}\sigma_0 \tag{4.31}$$

The quantities $\sigma_\perp$, $\sigma_\wedge$ are also called the Pedersen and Hall conductivity respectively. For a nonmagnetic plasma we recover from (4.31) the isotropic result

$$\sigma_\perp = \sigma_\parallel = \sigma_0, \quad \sigma_\wedge = 0$$

For weak magnetic fields $\overline{\nu}_{e,i} >> \omega_{c,e}$, (4.31) tells us that

$$\sigma_\wedge << \sigma_\perp \approx \sigma_\parallel$$

so that the conductivity is almost fully isotropic. For strong fields $\overline{\nu}_{e,i} << \omega_{c,e}$, (4.31) implies that

$$\sigma_\perp << \sigma_\wedge << \sigma_\parallel$$

When $\overline{\nu}_{e,i} = \omega_{c,e}$ it follows that

$$\sigma_\perp = \sigma_\wedge = \frac{1}{2}\sigma_\parallel$$

The conductivity tensor σ has lateral isotropy with respect to the z-axis, hence with respect to the magnetic field. In the presence of a magnetic field all second-order phenomological tensors have lateral isotropy with respect to the magnetic field and have the same structure as (4.30). Incidentally, the present analysis reveals that the generalized Ohm's law is obtained from a simplified version of the equation of motion of the electron fluid.

The equation for internal energy

In the equation for internal energy we have two terms for energy losses or gains: the heat flux and the Ohmic dissipation. There is no loss term due to radiation because we have forgotten about radiation all together. In the solar corona the energy loss function, L, consists of three main terms, energy losses (or gains) through thermal conduction, energy losses through optically thin plasma radiation and energy gained through heating. The effect of the flux of heat is to reduce and eventually to wipe out spatial differences in temperature. It is common practice to put $\vec{\Phi}$ proportional to ∇T. The proportionality factor is really a tensor since the motion of charged particles along the field lines is rather different from that in planes normal to the magnetic field lines when the magnetic field is strong. Thus, L can be written as

$$\begin{aligned} L &= -\nabla.\vec{\Phi} + \rho^2 Q(T) - H \\ \vec{\Phi} &= \kappa \nabla T \end{aligned} \tag{4.32}$$

In the thermal conduction, κ is the anisotropic thermal conductivity tensor. In a strong magnetic field, the conduction across the magnetic field is inhibited whereas conduction along the field is unaffected. The thermal conductivity tensor has the same structure as the electrical conductivity tensor (4.30). It is characterized by three scalar quantities $\kappa_\parallel, \kappa_\perp, \kappa_\wedge$. These quantities satisfy the same type of inequalities as their corresponding electric conductivity quantities. The important quantity is $\kappa_\parallel$. It is given, to a good approximation, by

$$\kappa_\parallel = 10^{-11} T_\star^{5/2} \mathrm{W\ m^{-1} K^{-1}}$$

where $T_\star$ is the numerical value of temperature expressed in K. $Q(T)$ is the optically thin radiative loss function that has been observed and calculated. It may be approximated by a piecewise continuous function of the form

$$Q(T) = \chi T^\alpha$$

where χ and α vary between temperature ranges. Optically thin radiation is a valid assumption above a temperature of 10^4 K. H is the heating term that in the corona may consist of several terms. For example we may have

$$H = H_0 + H_\nu$$

where H_0 is the coronal heating function that depends on the coronal magnetic field, small-scale reconnection events and wave heating. H_ν is the viscous heating term. Remember also that $\mid \vec{j} \mid^2 / \sigma$ is the Ohmic heating term. L severely complicates the energy equation. In many applications we can neglect all these terms. So here is another major simplification of ideal and resistive MHD! Take

$$L = 0$$

An analysis based on typical length and time scales can be used to make this simplification plausible. The equation for internal energy is then

$$\begin{aligned} \frac{dp}{dt} - \frac{\gamma p}{\rho}\frac{d\rho}{dt} &= 0 \quad \text{ideal MHD} \\ &= (\gamma - 1)\frac{\mid \vec{j} \mid^2}{\sigma} \quad \text{resistive MHD} \qquad (4.33) \end{aligned}$$

The induction equation

It turns out that we do not need all of the pre-Maxwell equations (4.23). The third of these equations implies that

$$\nabla . \vec{j} = 0$$

This seems to be in contradiction with the equation for charge conservation. Recall that we obtained the third pre-Maxwell equation by dropping the displacement current as a $(v/c)^2$ contribution to the original third Maxwell equation. The interpretation of $\nabla . \vec{j} = 0$ is that $\nabla . \vec{j}$ is proportional to $(v/c)^2$, not that it is exactly 0. The equation for charge conservation is an equality between two quantities that are both proportional to $(v/c)^2$. In a non-relativistic theory this equation can be thrown over board.

We can use the 4th Maxwell equation for computing Q, or better we can note that Q does not appear in any of the remaining equations and drop the fourth of the Maxwell equations from our system of equations.

Let us now look at the electric field $\vec{E}$ and realize that it only appears in the first pre-Maxwell equation and in the generalized Ohm's law. Hence, we can combine these two equations into an evolution equation for $\vec{B}$ which is known as the induction equation. When we do that using the ideal or resistive approximation of Ohm's law we obtain:

$$\begin{aligned} \frac{\partial \vec{B}}{\partial t} &= \nabla \times \left(\vec{v} \times \vec{B}\right) \quad \text{ideal MHD} \\ &= \nabla \times \left(\vec{v} \times \vec{B}\right) + \eta \nabla^2 \vec{B} \quad \text{resistive MHD} \end{aligned} \tag{4.34}$$

Here $\eta = 1/(\mu\sigma)$ is the magnetic diffusivity (4.12) (which we have assumed to be spatially constant). Note that $\eta \neq \bar{\eta}$! The second of the pre-Maxwell equations (4.23) is then an initial condition for $\vec{B}$. Finally we can eliminate $\vec{j}$ from the equation of motion and the energy equation by using the third of the pre-Maxwell equations (4.23) $\vec{j} = (\nabla \times \vec{B})/\mu$.

Summary

The closed systems of equations of ideal and resistive MHD are

mass

$$\frac{\partial \rho}{\partial t} = -\nabla.(\rho \vec{v})$$

momentum

$$\rho \frac{d\vec{v}}{dt} = -\nabla p + \rho \vec{g} + \frac{1}{\mu}(\nabla \times \vec{B}) \times \vec{B}$$

internal energy

$$\frac{dp}{dt} = -\gamma p \nabla.\vec{v} + \underbrace{(\gamma - 1)\frac{|\nabla \times \vec{B}|^2}{\mu^2 \sigma}}$$

induction

$$\frac{\partial \vec{B}}{\partial t} = \nabla \times \left(\vec{v} \times \vec{B}\right) + \underbrace{\eta \nabla^2 \vec{B}}$$

$$\nabla.\vec{B} = 0 \tag{4.35}$$

The resistive terms are underbraced for clarity. The electric current density $\vec{j}$ and the electric field $\vec{E}$ have been eliminated by the use of

$$\vec{j} = \frac{1}{\mu}(\nabla \times \vec{B}), \quad \vec{E} = -\vec{v} \times \vec{B} + \underbrace{\eta(\nabla \times \vec{B})}$$

The equations (4.35) are written with both local partial time derivatives $\partial/\partial t$ and convective time derivatives d/dt. Rewriting these equations in fully Eulerian form with only local partial time derivatives $\partial/\partial t$ or in fully Lagrangian form with only convective time derivatives d/dt is straightforward.

Conditions of validity

This is a good point to look back and reflect on the assumptions and restrictions that have been made in order to arrive at the MHD equations (4.35). Let us recall that the Maxwell-Boltzman description (2.7) - (2.9) using distribution functions for the various species in phase space was our starting point. From that description we moved on to multi-fluid theory in order to finally arrive at Magnetohydrodynamics which is a single-fluid theory where the distinction between different species has been lost. By now the attentive reader has figured out that there is a hierarchy of temporal and spatial scales in a plasma. When we move from microscopic to macroscopic, the first level corresponds to the very high frequencies and very short length scales defined by the dynamics of the light electrons. The second level is concerned with the dynamics of the heavy ions. Here we have low frequencies and short length scales characterized by $\omega_{c,i}$ and $r_{L,i}$. The last level is that of the very-low-frequency long length scale fluid behaviour of plasmas. Hence, the approximation of MHD makes physical sense for plasma situations where the spatial and temporal scales of the variations of the fluids and fields are substantially longer than the corresponding scales of the heaviest component of the plasma, i.e. the ions. If ω and L are the characteristic frequency and the characteristic length scale of any variation then, in order for MHD to be a valid theory for its description, we must have

$$\boxed{\begin{aligned} &\frac{r_{L,i}}{L} << 1 \\ &\frac{\omega}{\omega_{c,i}} << 1 \qquad (4.36) \end{aligned}}$$

This condition has emerged on several occasions. In particular it is the condition (4.17), (4.18) for neglecting the Hall term compared to the induction term and to arrive at the ideal MHD approximation of the generalized Ohm's law. It is a vital condition for MHD to be a valid approximation of plasma behaviour.

The single-fluid equations of MHD assume that the collisions lead to locally nearly Maxwellian distribution functions for both electrons and ions so that the pressure tensors for electrons and ions have only contributions from the isotropic electron pressure and the isotropic ion pressure. For ions the dominant collision mechanism is ion-ion interactions, characterized by an averaged collision time $\tau_{i,i} = 1/\overline{\nu}_{i,i}$, with $\overline{\nu}_{i,i}$ the average frequency of collisions of ions with other ions. For electrons it is interactions with either ions or with other electrons. Since

$$\tau_{e,e} \approx \tau_{e,i}$$

we do not need to make the distinction. In order that sufficient collisions (high collisionality) take place to make the distribution functions locally nearly Maxwellian, it is required that the collision times are short compared to the global time scale of the system defined by e.g, the thermal velocity of the ions $v_{t,i}$ and the size L of the system. Hence

$$\frac{\tau_{e,i} v_{t,i}}{L} << 1$$
$$\frac{\tau_{i,i} v_{t,i}}{L} << 1 \qquad (4.37)$$

Since

$$\tau_{i,i} = (m_i/m_e)^{1/2} \tau_{e,i}$$

the second of inequalities (4.37) sets the stronger condition. The heavy ions take much longer than the light electrons to reach isotropy. (4.37) also imply that the mean free paths of the ions and electrons are much shorter that L. If (4.37) are satisfied, the viscosity is negligible

$$\mid \nabla . \mathbf{P_i} \mid / \mid \nabla p_i \mid << 1$$

The assumption (4.7) that the ions and electrons have the same temperature $T_i = T_e$ means that the energy equilibration time τ_{eq} is much shorter that global time scale $L/v_{t,i}$. It can be shown that

$$\tau_{eq} \approx (m_i/m_e) \tau_{e,i}$$

Hence,

$$\frac{m_i}{m_e} \frac{\tau_{e,i} v_{t,i}}{L} << 1 \qquad (4.38)$$

It can be shown that, when inequality (4.38) is satisfied, thermal conduction in the energy equation can be neglected. (4.38) is a stronger condition than (4.37), so that we only need to worry about (4.38). (4.36) and (4.38) are conditions for MHD to be an accurate representation of a plasma phenomenon characterized by a given temporal scale τ. Actually there are four possibilities.

- $\tau < \tau_{e,i}$. Both the electrons and the ions are kinetic in the sense that they require a kinetic theory for their description. T_e and T_i are kinetic temperatures. The continuum thermodynamic definition of temperature does not apply.

- $\tau_{e,i} < \tau < \tau_{i,i}$. The electrons behave as a fluid. The ions are kinetic and T_i is kinetic.

- $\tau_{i,i} < \tau < \tau_{eq}$. Both the electrons and the ions behave as a fluid. Hence a two-fluid model with isotropic electron and isotropic ion pressure is applicable. However, $T_e \neq T_i$.

- $\tau_{eq} < \tau$. One-fluid model with $T_e = T_i$ is all right.

Ideal MHD requires the additional condition that the magnetic Reynolds number R_m (4.11), is much larger than 1

$$R_m >> 1 \qquad (4.39)$$

The conditions for small gyro-radius (4.36), and large magnetic Reynolds number (4.39), are well satisfied for a wide range of plasmas of fusion and astrophysical interest. However, condition (4.38) for high collisionality is not satisfied for plasmas of fusion interest and probably also for many plasmas of astrophysical interest. This conclusion is in striking contradiction with the overwhelming evidence that MHD provides a very accurate description of most large-scale low-frequency plasma behaviour in tokamaks. The reason why MHD gives a more accurate description than anticipated on the basis of condition (4.38), is linked to the role of the magnetic field in collisionless plasmas. In a collisionless plasma, the magnetic field plays the role of collisions for motions perpendicular to the magnetic field lines. Perpendicular to the field lines the particles are confined to the vicinity of a field line and execute 2-dimensional nearly isotropic motions at least when their Larmor radius is much shorter than the size L of the system. The perpendicular motion is fluid-like. The perpendicular components of the equation of motion provide an excellent description of plasma behaviour in both collision-dominated and collisionless plasmas. In the direction parallel to the magnetic field, MHD treats the physics inaccurately. However, it turns out that in the collisionless regime, the part of dynamics that is inaccurately described by MHD, does not matter. As far as the energy equation is concerned, it turns out the collision-dominated and the collisionless equations are identical when the plasma motions are incompressible. Deviations do occur for compressible motions.

4.5 The induction equation and conservation of magnetic flux

The aim of the present Section is to show that the induction equation implies that the magnetic field lines are anchored to the plasma or in Alfvén's words "the matter of the fluid is fastened to the lines of forces." This anchoring of the magnetic fields to the plasma is perfect in ideal MHD, in resistive MHD there is a possible slippage of the magnetic field lines.

The important quantity in this Section is the total time derivative of the magnetic flux through a surface that moves with the plasma as a whole. From analysis of vector fields we know that the total time derivative of the flux of a vector $\vec{A}$ through a surface S that moves with velocity $\vec{u}$ is given by

$$\frac{d\Phi}{dt} = \frac{d}{dt}\int \vec{A}.\vec{n}\, dS = \int \left\{ \frac{\partial \vec{A}}{\partial t} - \nabla \times (\vec{u} \times \vec{A}) + \vec{u}\nabla.\vec{A} \right\} .\vec{n}\, dS$$

With $\nabla.\vec{B} = 0$, the total time derivative of the magnetic flux through a surface that moves with velocity $\vec{u}$ is

$$\frac{d}{dt}\int_S \vec{B}.\vec{n}dS = \int \left\{ \frac{\partial \vec{B}}{\partial t} - \nabla \times (\vec{u} \times \vec{B}) \right\} .\vec{n}\, dS$$

The time derivative of the magnetic field is given by the induction equation (4.34). Let us now consider a surface S that moves with the plasma as a whole (i.e. put $\vec{u} = \vec{v}$) and adopt the ideal MHD approximation of the induction equation for computing the time derivative of $\vec{B}$. The result is a *cornerstone equation* of ideal MHD

$$\frac{d}{dt}\int \vec{B}.\vec{n}\, dS = 0 \qquad (4.40)$$

This equation tells us that magnetic flux through any surface that moves with the plasma is invariant in time in ideal MHD or that the magnetic flux through any surface that moves with the plasma is conserved!

Flux conservation implies that the magnetic field lines are anchored to the plasma or that they are frozen into the plasma or that they are glued to the plasma. This can be easily understood by looking at a *magnetic flux tube.* A magnetic flux tube is defined as the cylindrical volume that is enclosed by the collection of field lines that intersect a given closed curve C. The name magnetic flux tube will also be used to indicate the cylindrical surface formed by the collection of field lines that intersect the closed curve C. A given magnetic flux tube contains a constant amount of magnetic flux in the sense that

$$\int_S \vec{B}.\vec{n}\, dS = \text{constant along the flux tube.}$$

This is a direct consequence of the definition of magnetic flux tube and the Maxwell equation that $\nabla.\vec{B} = 0$. There is no magnetic flux through any part of the cylindrical surface formed by the collection of field lines that intersect the closed curve C. We let this flux tube evolve in time and look at the plasma elements move with their local plasma velocity. Because of flux conservation plasma elements that are initially on the magnetic flux tube remain on the magnetic flux tube. *In ideal MHD the magnetic field evolves in such way as to preserve the integrity of each magnetic flux tube!*

Conservation of magnetic flux implies conservation of magnetic lines. Specify a given magnetic field line as the intersection of two magnetic flux tubes. Since plasma elements that are initially on a magnetic flux tube remain on that magnetic flux tube forever, it follows that plasma elements that are initially on the intersection of two magnetic flux tubes remain on that intersection, or plasma elements that are initially on a given magnetic field line remain on that magnetic field line forever. *The magnetic field lines are frozen into the plasma.* This is the basic property of ideal MHD! The magnetic field lines can be visualized as a spider web that is dragged along and deformed by the flow of the plasma. Whether the plasma motion controls the magnetic field or vice versa depends on the amount of energy in the flow and in the magnetic field respectively. In what follows we shall see that the magnetic field lines do not like to be deformed and produce a Lorentz force that opposes the motions that induce these deformations. Attaching the plasma to the magnetic field lines has the effect of giving mass to the field lines and this circumstance is seen clearly in shear Alfvén waves.

Deviations from ideal flux freezing occur due to effects that are absent in ideal MHD. To make this clear let us start from the generalized Ohm's law (4.9). As before we use the first and the third of the pre-Maxwell equations (4.23) to obtain

$$\begin{aligned}
\frac{\partial \vec{B}}{\partial t} &= \nabla \times (\vec{v} \times \vec{B}) - \nabla \times \left(\frac{1}{en_e}\vec{j} \times \vec{B}\right) - \nabla \times (\eta \nabla \times \vec{B}) + \nabla \times \left(\frac{\nabla p_e}{en_e}\right) \\
&= \nabla \times (\vec{v_e} \times \vec{B}) - \nabla \times (\eta \nabla \times \vec{B}) + \nabla \times \left(\frac{\nabla p_e}{en_e}\right)
\end{aligned}$$

This is the induction equation (4.34) but now we have included the Hall term and the battery term. Let us reiterate that the effect of the battery term is small from a global point of view but

$$\nabla \times \left(\frac{\nabla p_e}{en_e}\right) = \frac{1}{en_e^2} \nabla p_e \times \nabla n_e$$

so that the battery term causes an Eulerian temporal change of $\vec{B}$ if there is a gradient of n_e oblique to the gradient of p_e.

Let us now compute the change in time of the magnetic flux through a surface that moves with the velocity of the plasma $\vec{v}$ or with the velocity of the electron gas $\vec{v}_e$. The total time derivative of the magnetic flux through a surface that moves with the plasma as a whole is

$$\begin{aligned} \frac{d}{dt}\int_S \vec{B}.\vec{n}dS &= \int_S \nabla \times \left(-\eta\nabla \times \vec{B} - \frac{1}{en_e}\vec{j} \times \vec{B} + \frac{\nabla p_e}{en_e}\right).\vec{n}\, dS \\ &= \oint_C \left(-\eta\nabla \times \vec{B} - \frac{1}{en_e}\vec{j} \times \vec{B} + \frac{\nabla p_e}{en_e}\right).\vec{1}_t\, ds \\ &= -\oint_C \vec{E}^\star.\vec{1}_t\, ds = -\mathrm{EMF} \end{aligned}$$

with EMF the electromotive force per unit charge that is produced by $\vec{E}^\star$ around the contour C. This result is of course no surprise. From a formal point of view it is the standard relation between the temporal variation of the magnetic flux and the electromotive flux. However, our result is more than a formal relation since $\vec{E}^\star$ is given in terms of macroscopic plasma quantities. The conclusion is that if the freezing of magnetic flux to the plasma flow is violated on any given moment and at any given place that this is due to

$$\nabla \times \left(-\eta\nabla \times \vec{B} - \frac{1}{en_e}\vec{j} \times \vec{B} + \frac{\nabla p_e}{en_e}\right)$$

This term is responsible for the fact that the magnetic field lines are not perfectly glued to the plasma and that they can slip with respect to the plasma.

We have already noted that the Hall term gives the largest contribution to the electric field $\vec{E}^\star = \vec{E} + \vec{v} \times \vec{B}$. The total time derivative of the magnetic flux through a surface that moves with the electron gas is

$$\begin{aligned} \frac{d}{dt}\int_{S_e} \vec{B}.\vec{n}dS_e &= \int_{S_e} \nabla \times \left(-\eta\nabla \times \vec{B} + \frac{\nabla p_e}{en_e}\right).\vec{n}\, dS_e \\ &= \oint_{C_e} \left(-\eta\nabla \times \vec{B} + \frac{\nabla p_e}{en_e}\right).\vec{1}_t\, ds \\ &= \oint_{C_e} \vec{E}^\star_e.\vec{1}_t\, ds = -\mathrm{EMF}_e \end{aligned}$$

Conservation of magnetic flux through a surface that moves with the electron gas is a better approximation of reality than conservation of magnetic flux through a surface that moves with the plasma as a whole. This is because of the simple fact that the heavy ions are much less well glued to the magnetic field lines than the light electrons. Deviation from conservation of magnetic flux through a surface that moves with the electron gas is due to the Ohmic term and to the battery term. These two terms are responsible for the fact that the magnetic field lines are not perfectly glued to the electrons and that they can slip with respect to the electrons.

4.6 The diffusive limit of the induction equation

This paragraph is concerned with the induction equation and in particular with the effect of the Ohmic term. We go back to the induction equation in resistive MHD (4.34) and remember that we have dropped the resistive term $\eta\nabla^2\vec{B}$ with the argument that the magnetic Reynolds number $R_m \gg 1$. Neglecting the resistive term in the induction equation lowers the order of the induction equation seen as a differential equation. This is usually a good indication of the existence of boundary layers.

In the resistive version of the induction equation the first term on the right hand side describes the fact that the magnetic field lines are glued to the plasma. Let us now determine the effect of the second term. In order to do so we consider the limiting case $R_m \ll 1$ so that we can drop the convective term. The induction equation then takes the form of a diffusion equation

$$\frac{\partial \vec{B}}{\partial t} = \eta \, \nabla^2 \vec{B} \qquad (4.41)$$

Let us consider the simple case of a unidirectional magnetic field that varies in only one direction perpendicular to the magnetic field lines in a system of Cartesian coordinates, for example

$$\vec{B}(x, y, z; t) = B(x; t)\vec{1}_z.$$

For this magnetic field the induction equation takes the form

$$\frac{\partial B}{\partial t} = \eta \frac{\partial^2 B}{\partial x^2}$$

Look now at a position $x = x_{max}$ where B_z attains a local maximum. At this position

$$\frac{\partial^2 B}{\partial x^2} < 0 \quad \rightarrow \quad \frac{\partial \vec{B}}{\partial t} < 0$$

so that B_z decreases there in time. Similarly, at a position $x = x_{min}$ where B_z attains a local minimum

$$\frac{\partial^2 B}{\partial x^2} > 0 \quad \rightarrow \quad \frac{\partial \vec{B}}{\partial t} > 0$$

so that B_z increases there in time. Thus the term we are looking at tends to smooth out any inhomogeneity in B_z and aims to reduce B_z to a constant value in the whole space. In other words the magnetic field diffuses from regions of high B to low field regions. In this η plays the role of a diffusion coefficient. The typical time scale of this diffusion process τ_η can be estimated as

$$\frac{B}{\tau_\eta} = \eta \frac{B}{L^2}$$

so that

$$\tau_\eta = \frac{L^2}{\eta} \qquad (4.42)$$

The electromagnetic diffusion time scale τ_η can be used to rewrite the magnetic Reynolds number as the ratio of two time scales since

$$R_m = \frac{Lv}{\eta} = \frac{\tau_\eta}{L/v} = \frac{\tau_\eta}{\tau_d} \qquad (4.43)$$

with τ_d the dynamical time scale. If we take the Alfvén velocity v_A for v, then the dynamical time scale is the Alfvén transit time $\tau_A = \frac{L}{v_A}$ and R_m is then known as the Lundquist number Lu. The Lundquist number is the ratio of the electromagnetic diffusion time scale τ_η to the Alfvén transit time τ_A

$$Lu = \frac{\tau_\eta}{\tau_A} \qquad (4.44)$$

The condition $Lu \gg 1$ means that the characteristic time scale of the diffusion of the magnetic field is very long compared with the dynamical time scale and that for phenomena with time scales comparable to τ_A the diffusion of the magnetic field is unimportant. Conversely, the condition $Lu \ll 1$ means that the characteristic time scale of the diffusion of the magnetic field is very short compared with the dynamical time scale so that the induction term $\nabla \times (\vec{v} \times \vec{B})$ can be ignored for the evolution of $\vec{B}$.

The condition $Lu \gg 1$ is globally almost always satisfied, but local violations of this condition can and actually do often occur. An electrical current always experiences resistance from the plasma in which it flows. The ideal situation $Lu = \infty$ obviously does not occur in nature. For time spans that are short compared to τ_η, the plasma behaves (to a very approximation) as if it is a perfect conductor and the magnetic field lines follow the motion of the plasma. A temporal change in $\vec{B}$ is mainly due to a redistribution of the magnetic field lines by the motion of the plasma. However, this picture does not hold for time spans that are comparable to τ_η. The magnetic field lines do not follow the motion of the plasma and the magnetic flux in a material volume can change substantially. Ohmic diffusion provides an important contribution to the temporal change in $\vec{B}$.

The discussion becomes even more complicated if you go back to the definition of τ_η and note that the velocity field $\vec{v}$ can be nonuniform and that L is the length scale for the spatial variation of the magnetic field. L is determined by the gradients of the components of $\vec{B}$. These gradients can undergo large variations over the whole plasma and can be small in some parts of the plasma and large elsewhere. Electromagnetic diffusion is important at the positions where the gradients of the components of $\vec{B}$ are large and relatively unimportant where these gradients are small. Strictly speaking we should treat τ_η as a function of position. Local deviations from the approximation of an ideal conductor occur at positions where the local value of τ_η is of the same order as or small compared to the Alfvén transit time τ_A.

The magnetic Reynolds number or the Lundquist number tells us whether the temporal evolution of the magnetic field is determined by the flow of the plasma or by diffusion. When $Lu \gg 1$ the diffusion term in the induction term can be neglected and the temporal evolution of the magnetic field is determined by the flow. The magnetic field is frozen into the plasma. The solar wind is a good example of a plasma with a high Lundquist number: $Lu \approx 10^{16}-10^{17}$. As a consequence there is almost no diffusion of the magnetic field in the solar wind. Only the component of velocity perpendicular to $\vec{B}$ enters in the convective term. Any flow along the field lines has no consequences. When $Lu \approx 1$, diffusion is important for the temporal evolution of the magnetic field and can dominate this evolution. Also in a diffusion dominated region, plasma can almost freely flow across the magnetic field lines without any substantial effect on the magnetic field.

The evolution of the magnetic field is completely determined by resistive effects at positions where

$$\nabla \times (\vec{v} \times \vec{B}) = 0$$

The magnetic field changes only on the long electromagnetic diffusion time scale τ_η. On shorter time spans the magnetic field can be considered to be time independent. In ideal MHD the previous equation is the condition for the magnetic field to be stationary or time independent.

Since the resistive term $\eta \nabla^2 \vec{B}$ is only important in thin layers where either the magnetic field is characterized by large gradients and/or $\nabla \times (\vec{v} \times \vec{B}) \approx 0$ the term resistive boundary layers is used in analogy with the well-known viscous boundary layers of fluid dynamics.

4.7 Magnetic field lines

In this Section we shall first look at the mathematical representation of magnetic field lines. This only involves repetition of simple concepts of differential geometry of curves. Then we shall look at an important consequence of the fact that in ideal MHD the magnetic field lines are glued to the plasma or frozen into the plasma.

A 3-dimensional curve can be expressed in parametric form as

$$x = f_1(u), \quad y = f_2(u), \quad z = f_3(u)$$

in terms of a parameter u. $f_1(u), f_2(u), f_3(u)$ are three known functions of the parameter u, they can be seen as the three components of the vector function $\vec{f}(u)$. The tangent vector to this curve is

$$\vec{t} = \frac{d\vec{x}(u)}{du} = [\frac{dx(u)}{du}, \frac{dy(u)}{du}, \frac{dz(u)}{du}]^t$$

The magnetic field lines are the 3-dimensional curves with the magnetic field being tangent to this curve in all points. So for a magnetic field

$$\vec{B} = [B_x, B_y, B_z]^t$$

the differential equations for the magnetic field lines are

$$\frac{dx(u)}{du} = \lambda B_x, \quad \frac{dy(u)}{du} = \lambda B_y, \quad \frac{dz(u)}{du} = \lambda B_z$$

The factor of proportionality λ can be a function of position and time. It is straightforward to show that

$$\lambda = \frac{1}{B}\frac{ds}{du}$$

with s the arc length along the magnetic field line and B the strength of the magnetic field

$$\begin{aligned} (\frac{ds}{du})^2 &= (\frac{dx}{du})^2 + (\frac{dy}{du})^2 + (\frac{dz}{du})^2 \\ B &= (B_x^2 + B_y^2 + B_z^2)^{1/2} \end{aligned}$$

With the arc length s as parameter, the differential equations for the magnetic field lines are

$$\frac{dx(s)}{ds} = \frac{B_x}{B}, \quad \frac{dy(s)}{ds} = \frac{B_y}{B}, \quad \frac{dz(s)}{ds} = \frac{B_z}{B} \tag{4.45}$$

Let us now look at a 2-dimensional magnetic field in Cartesian coordinates (actually it is a poloidal magnetic field with invariance in the $y-$direction)

$$B_x = B_x(x, z), \quad B_y = 0, \quad B_z = B_z(x, z)$$

The magnetic field lines are now planar curves in the $xz-$plane (or better in planes parallel to the $xz-$plane). The differential equations for the magnetic field lines are

$$\frac{dx(s)}{ds} = \frac{B_x}{B}, \quad \frac{dz(s)}{ds} = \frac{B_z}{B}$$

and can be combined into one equation

$$\frac{dz}{dx} = \frac{B_z}{B_x}$$

This is of course a familiar result. It is the classic first order differential equation for 2-dimensional curves with prescribed direction field. A 2-dimensional curve can be seen as the graphic representation of an equation

$$\Psi(x, z) = C$$

Along that curve

$$\frac{\partial \Psi}{\partial x} + \frac{\partial \Psi}{\partial z}\frac{dz}{dx} = 0$$

so that

$$\frac{dz}{dx} = -\frac{\partial \Psi / \partial x}{\partial \Psi / \partial z}$$

When we apply this result to the field lines of our poloidal $y-$invariant magnetic field, it follows that

$$B_x = -\frac{\partial \Psi}{\partial z}, \quad B_z = \frac{\partial \Psi}{\partial x}$$

so that

$$B_x \frac{\partial \Psi}{\partial x} + B_z \frac{\partial \Psi}{\partial z} = (\vec{B}.\nabla)\Psi = 0.$$

The derivative of Ψ along the magnetic field line is zero (this is not a big surprise since $\Psi = C$ is the equation of the magnetic field line). Ψ is the stream function of the poloidal magnetic field. Because of the adopted y-invariance the equation

$$\Psi(x, z) = C$$

represents a surface in 3 dimensions. It is the magnetic surface associated with the constant C.

For our poloidal $y-$invariant magnetic field we can trivially satisfy the condition $\nabla.\vec{B} = 0$ by writing

$$\vec{B} = \nabla \times (\Psi(x,z)\vec{1}_y) \qquad (4.46)$$

There is no room for confusion, this is indeed the stream line function of the magnetic field.

The reader knows from his/her first year undergraduate course on elementary differential geometry that the vectorial velocity and the vectorial acceleration of a parametrized 3-dimensional curve are given by

$$\frac{d\vec{x}}{du} = \frac{ds}{du}\vec{1}_t, \quad \frac{d^2\vec{x}}{du^2}$$

respectively, where $\vec{1}_t$ is the unit vector tangent to the curve. For a magnetic field line $\vec{1}_t = \vec{1}_B$, the vectorial velocity and the vectorial acceleration span the osculating plane. For our 2-dimensional $y-$invariant magnetic field these are two vectors in the $xz-$plane and the osculating plane is simply $y = Cte$. The unit vector normal to the osculating plane is $\vec{1}_y$. Of course $\vec{1}_B$ is the unit vector in the magnetic surface and parallel to the magnetic field lines and $\vec{1}_y$ is the unit vector in the magnetic surface and perpendicular to the magnetic field lines. So we have two orthonormal directions in the magnetic surfaces; the third orthonormal direction is defined by the unit vector normal to the magnetic surfaces.

Let us now see what it means that the magnetic field lines are glued to the plasma for our 2-dimensional $y-$invariant poloidal magnetic field. In particular, we are interested in understanding what happens when there is a velocity field that is

• incompressible (no compression or expansion of the plasma)
• in the magnetic surfaces
• perpendicular to the magnetic field lines!

Hence

$$\vec{v}(x,z) = v_y(x,z)\vec{1}_y$$

We can write the induction equation as

$$\frac{\partial \vec{B}}{\partial t} = (\vec{B}.\nabla)v_y\vec{1}_y = B\frac{dv_y}{ds}\vec{1}_y \qquad (4.47)$$

This incompressible flow in the magnetic surfaces and perpendicular to the magnetic field lines creates a time dependent magnetic field in the $y-$direction. This time dependent magnetic field in the $y-$direction tries to counteract the velocity field that deforms the magnetic field lines. A necessary condition for a stationary plasma is that $\frac{\partial \vec{B}}{\partial t} = 0$ which leads to the following condition for v_y

$$(\vec{B}.\nabla)v_y = B\frac{dv_y}{ds} = 0$$

so that v_y is constant on a magnetic field line or

$$v_y = v_y(\Psi) \qquad (4.48)$$

Nonuniformity of v_y along a poloidal magnetic field line would lift the field line out of its osculating plane and would generate a time dependent magnetic field in the $y-$direction. The field lines have to be displaced without being deformed.

Let us now look at an axisymmetric poloidal magnetic field in cylindrical coordinates (r, φ, z)

$$\vec{B}_p = B_r(r,z)\vec{1}_r + B_z(r,z)\vec{1}_z$$

For our poloidal $\varphi-$invariant magnetic field we can trivially satisfy the condition $\nabla.\vec{B} = 0$ by writing

$$\boxed{\vec{B}_p = \nabla \times (\Psi(r,z)\frac{\vec{1}_\varphi}{r}) \qquad (4.49)}$$

Since $\nabla \times (\frac{\vec{1}_\varphi}{r}) = 0$ we can write $\vec{B}_p$ as

$$\vec{B}_p = \nabla\Psi(r,z) \times \frac{\vec{1}_\varphi}{r} = -\frac{1}{r}\frac{\partial\Psi}{\partial z}\vec{1}_r + \frac{1}{r}\frac{\partial\Psi}{\partial r}\vec{1}_z$$

Hence

$$B_r\frac{\partial\Psi}{\partial r} + B_z\frac{\partial\Psi}{\partial z} = (B_p.\nabla)\Psi = 0$$

This means that $\Psi(r,z)$ is constant along a poloidal magnetic field line. For this reason $\Psi(r,z)$ is named the stream function of the poloidal magnetic field B_p. The equation of a poloidal magnetic field line is

$$\Psi(r,z) = C$$

Actually, because of the axisymmetry $(\partial/\partial\varphi = 0)$ this is the equation of a magnetic surface.

Let us look at a particularly simple situation of a straight magnetic field

$$\boxed{\vec{B}_p = B_z(r)\vec{1}_z \qquad (4.50)}$$

Since

$$B_r = -\frac{1}{r}\frac{\partial\Psi}{\partial z} = 0$$

it follows that

$$\Psi = \Psi(r)$$

The magnetic surfaces $\Psi = C$ are cylinders. The unit vectors in the magnetic surfaces, respectively parallel and perpendicular to the magnetic field lines are $\vec{1}_z$ and $\vec{1}_\varphi$. The third orthogonal unit vector is $\vec{1}_r$.

Let us now see what it means that the magnetic field lines are glued to the plasma for this simple straight magnetic field. Again we are interested in a velocity field that is

- incompressible
- in the magnetic surfaces
- perpendicular to the magnetic field lines!

Hence

$$\vec{v}(r,z) = v_\varphi(r,z)\vec{1}_\varphi$$

Let us rewrite $v_\varphi(r,z)$ as

$$v_\varphi(r,z) = r\Omega(r,z)$$

and relate the azimuthal velocity field to a (possibly differential) rotation around the $z-$axis. It is straightforward to rewrite the induction equation as

$$\frac{\partial \vec{B}}{\partial t} = B_z \frac{\partial v_\varphi}{\partial z} \vec{1}_\varphi \qquad (4.51)$$

We can go back to (4.47) and repeat the observation we made there with y replaced with φ. The incompressible flow in the magnetic surfaces and perpendicular to the magnetic field lines creates a time dependent magnetic field in the φ−direction. This time dependent magnetic field in the φ−direction tries to counteract the velocity field that deforms the magnetic field lines. A necessary condition for a stationary plasma is that

$$\frac{\partial v_\varphi}{\partial z} = 0, \quad v_\varphi = v_\varphi(r) \qquad (4.52)$$

so that v_φ is constant on a magnetic field line. Nonuniformity of v_φ along a poloidal magnetic field line would lift the field line out of its osculating plane and would generate a time dependent magnetic field in the φ−direction. The field lines have to be displaced without being deformed.

Finally, let us look at an axisymmetric 2-dimensional poloidal magnetic field in cylindrical coordinates when there is an azimuthal velocity field

$$\vec{v}(r, z) = v_\varphi(r, z)\vec{1}_\varphi = r\Omega(r, z)\vec{1}_\varphi.$$

We have $\vec{v} \times \vec{B} = \Omega \nabla \Psi$ and

$$\frac{\partial \vec{B}}{\partial t} = \nabla\Omega \times \nabla\Psi = \frac{D(\Psi, \Omega)}{D(r, z)} \vec{1}_\varphi \qquad (4.53)$$

This is basically the same result as obtained in (4.47) and in (4.51). The plasma is stationary if

$$\nabla\Omega \times \nabla\Psi = 0$$

which means that $\nabla\Omega$ and $\nabla\Psi$ are parallel so that

$$\Omega = \alpha(\Psi) \qquad (4.54)$$

A necessary condition for a stationary equilibrium is that Ω is only a function of Ψ. Since Ψ is constant on a poloidal magnetic field line, this means that Ω is also constant on a poloidal magnetic field line. This is *Ferraro's isorotation law.* Nonuniformity of Ω along a poloidal magnetic field line would lift the field line out of its osculating plane and would generate a time dependent azimuthal magnetic field. The field lines have to be displaced without being deformed. Ferraro's isorotation law is a necessary condition for a stationary equilibrium. It implies that

$$v_\varphi = r\alpha(\Psi)$$

The magnetic fields (4.46), (4.49) and (4.50) have been subjected to incompressible motions that are in the magnetic surfaces and perpendicular to the magnetic field lines. This resulted in the generation of time dependent magnetic fields (4.47), (4.51) and (4.53) counteracting the motions that deform the magnetic field lines. Stationarity requires that the motions displace the magnetic lines without deformation as is dictated by (4.48), (4.52) and (4.54).

4.8 The Lorentz force

In the previous Section we have looked at the kinematics of the magnetic field when a class of specific motions was prescribed. The motions were incompressible and directed in the magnetic surfaces perpendicular to the magnetic field lines. We have seen that in order to have a stationary situation the motions should not deform the magnetic field lines. Obviously, that is not the end of the story. In order to understand the reaction of the magnetic field we need to look at the Lorentz force. The Lorentz force is

$$\vec{j} \times \vec{B} = \frac{1}{\mu}(\nabla \times \vec{B}) \times \vec{B} = \frac{1}{\mu}(\vec{B}.\nabla)\vec{B} - \nabla(\frac{B^2}{2\mu})$$

$(\vec{B}.\nabla)$ is the directional derivative along the magnetic field line. Hence $(\vec{B}.\nabla) = B\frac{d}{ds}$ with s the arc length along the magnetic field line. We denote the unit vector tangential to the magnetic field line as $\vec{1}_t$ so that $\vec{1}_t = \frac{\vec{B}}{\| \vec{B} \|}$. The first term in the right hand side of the Lorentz force can be rewritten as

$$\frac{1}{\mu}(\vec{B}.\nabla)\vec{B} = \frac{B}{\mu}\frac{d}{ds}(B\vec{1}_t) = \frac{d}{ds}\left(\frac{B^2}{2\mu}\right)\vec{1}_t + \frac{B^2}{\mu}\frac{d\vec{1}_t}{ds}$$

At this point you have to go back to your undergraduate elementary course on differential geometry. Recall that for a 3-dimensional curve you can define three natural orthogonal directions: the tangent, the normal and the binormal and also recall the first formula of Frenet which specifies the variation of the unit vector tangent to the curve as you move along the curve:

$$\frac{d\vec{1}_t}{ds} = \frac{\vec{1}_n}{R_c}$$

R_c is the local radius of curvature and $\vec{1}_n$ the unit vector along the normal directed towards the local centre of curvature. Hence

$$\frac{1}{\mu}(\vec{B}.\nabla)\vec{B} = \frac{d}{ds}\left(\frac{B^2}{2\mu}\right)\vec{1}_t + \frac{B^2}{\mu}\frac{\vec{1}_n}{R_c}$$

Since you now know that there are three natural orthogonal directions associated with the magnetic field line you immediately understand that it makes sense to decompose ∇ as

$$-\nabla(\frac{B^2}{2\mu}) = -\frac{d}{ds}(\frac{B^2}{2\mu})\vec{1}_t - \nabla_\perp(\frac{B^2}{2\mu})$$

where $\nabla_\perp$ is the $\nabla-$ operator in the planes normal to the magnetic field lines. Combine these results to write the Lorentz force as

$$\vec{j} \times \vec{B} = -\nabla_\perp(\frac{B^2}{2\mu}) + \frac{B^2}{\mu R_c}\vec{1}_n \qquad (4.55)$$

The Lorentz force is confined to planes normal to the magnetic field lines and has no component along the magnetic field. This we already knew as $(\nabla \times \vec{B}) \times \vec{B}$ is perpendicular to both $\vec{B}$ and $\nabla \times \vec{B}$.

The first term in the Lorentz force,

$$-\nabla_{\perp}(\frac{B^2}{2\mu}) \qquad (4.56)$$

is the *magnetic pressure force*. It is isotropic in the planes normal to the magnetic field lines and is directed from high magnetic pressure (high field strength) to low magnetic pressure (low field strength) in the same way as the gas pressure. The magnetic pressure force is present when the strength of the magnetic field is non-constant in space.

The second term in the Lorentz force,

$$\frac{B^2}{\mu R_c}\vec{1}_n \qquad (4.57)$$

is the *magnetic tension force*, which we shall denote as $\vec{T}_n$ in what follows. It is directed to the local centre of curvature and its magnitude is inversely proportional to R_c so that the more a field line is curved the stronger the tension force is. This is no surprise. Everybody who has played with a sling-shot as a child (or also at a later age) knows that the tension force is directed to the local centre of curvature and that its magnitude can be increased by increasing the curvature of the sling-shot. This magnetic tension force behaves in an identical way as the tension force in a string. It is present for magnetic fields with curved field lines and it aims to make curved magnetic field lines straight.

The attentive reader has realized that magnetic tension gives us the answer to our question about what happens when Ferraro's law of isorotation is violated. Deformation of the magnetic field lines generates a magnetic tension force which is directed to the local centre of curvature and tries to make the magnetic field lines straight again. Recall from elementary physics that a transversal displacement of part of an elastic string generates transversal waves that propagate along the string. Since magnetic tension is similar to tension in an elastic string, a transversal displacement of part of the magnetic field line generates a transversal wave that propagates along the magnetic field line. This wave was first discovered by Alfvén and is named the *Alfvén wave*. Let us take this analogy between a string and a magnetic field line a step further. The velocity of the transversal waves propagating along an elastic string with tension T per unit length and mass m per unit length is

$$v = \left(\frac{T}{m}\right)^{1/2}$$

The magnetic tension and mass of a flux tube per unit length and per unit transversal area are respectively B^2/μ and ρ. The analogy with the elastic string suggests us to predict transversal waves that propagate along the magnetic field lines with the (Alfvén) velocity

$$\left(\frac{B^2}{\mu\rho}\right)^{1/2} = \frac{B}{\sqrt{\mu\rho}} = v_A = \text{Alfvén velocity} \qquad (4.58)$$

As anticipated in Section 4 of the present Chapter the Alfvén velocity v_A is indeed a fundamental quantity for the characterization of the very-low-frequency fluid-like dynamics of the magnetic field lines to which the plasma is fastened.

4.9 Recapitulation

"Every time I learn something new it pushes some old stuff out of my brain!"
Homer Simpson
The Simpsons

- At the end of the previous Chapter we came to the conclusion that our system of single-fluid equations is not closed. Closure requires an evolution equation for the electric current density $\vec{j}$ and expressions for the transfer of momentum due to collisions, the stress tensor and the flux of heat. The evolution equation for $\vec{j}$ is obtained by computing the first order charge moment of the Boltzmann equation and is known as the *generalized Ohm's law*. It contains the transfer of momentum of electrons to ions which was computed by making the assumption that this transfer is proportional to the relative velocity of the electrons to the ions.

- The generalized Ohm's law is further simplified by comparing the various terms with the induction term. The ideal approximation of Ohm's law only keeps the induction term and drops all other terms. This is the very low-frequency large-scale approximation taken to the extreme. The magnetic field changes in time because the plasma moves. The essential condition for this approximation to make physical sense is that the length scale of the plasma phenomenon under consideration is much longer than the ion Larmor radius.

- The Maxwell equations are simplified to the pre-Maxwell equations applicable to non-relativistic plasmas. The electromagnetic waves are removed from the scene.

- Classic MHD assumes isotropic electron and ion pressures and takes the temperatures of electrons and ions to be equal. All contributions of the electric field are dropped. The equation of internal energy contains several complicated terms due to Ohmic dissipation, radiation, conduction and heating. In ideal MHD all these terms are neglected.

- Combination of Ohm's law and the 3-rd pre-Maxwell equation gives the evolution equation for the magnetic field, known as the induction equation. In ideal MHD this induction equation implies that magnetic flux through any surface that moves with the velocity of the plasma is conserved and also that the magnetic field lines are frozen into the plasma or that the plasma is glued to the magnetic field lines. The property of flux conservation and frozen-in magnetic field lines is the most basic property of ideal MHD!

- Deviation from flux conservation occurs because of terms other than the induction term. Inclusion of the Hall term shows that the electron fluid is the only component that you can be sure about to be glued to the magnetic field lines. Deviation from flux conservation also occurs because of resistive effects. The resistive term makes the magnetic field diffuse away when the magnetic field is spatially inhomogeneous. This process of diffusion takes place on a (long) time scale. The Lundquist number tells us whether the temporal evolution of the magnetic field is dictated by the plasma flow ($Lu \gg 1$) or by diffusion ($Lu \approx 1$).

- Incompressible motions in the magnetic flux surfaces perpendicular to the magnetic field lines generate a time dependent magnetic field in that direction unless the magnetic field lines are displaced without being deformed.
- The Lorentz force can be written as the sum of a magnetic pressure force and a tension force. The magnetic pressure force is isotropic in planes normal to the magnetic field lines and is directed from high to low pressure in the same way as the gas pressure. The magnetic tension force is present for magnetic fields with curved field lines. It is directed to the local centre of curvature and its magnitude is inversely proportional to the local radius of curvature.
- Since magnetic field lines are loaded with plasma, the magnetic tension force will cause transversal waves. These transversal waves are predicted to propagate along the magnetic field lines at the Alfvén velocity, v_A. Hence, the Alfvén velocity v_A is a fundamental quantity for the characterization of the very-low-frequency fluid-like dynamics of the magnetic field lines to which the plasma is fastened.

4.10 Problems

"Now look", she said. "I'm sure there isn't a man living who wasn't a little scared his first time."
"Do you really believe I've never done this thing before?"
"Well," she said, "It's pretty obvious you haven't. You don't have the slightest idea what to do. You're nervous and awkward. You can't even-"
"Oh my God," Benjamin said.
"I mean just because you might be inadequate in one way doesn't-"
"Inadequate?!"
Conversation of Benjamin Braddock with Mrs. Robinson
The graduate
Charles Webb 1963

1. Show that

$$\frac{1}{en_e} = \frac{B/\rho}{\omega_{c,i}}$$

$$\frac{|\vec{j}\,|}{en_e} \approx \frac{r_{L,A,i}}{L} v_A$$

$$\begin{aligned}
\mid \vec{v}_e - \vec{v} \mid &\approx \frac{r_{L,A,i}}{L} v_A \\
\mid \vec{v}_i - \vec{v} \mid &\approx \frac{r_{L,A,i}}{L} \frac{m_e}{m_i} v_A \\
\frac{\rho_e \mid \vec{v}_e - \vec{v} \mid^2}{p_i} &\approx \frac{m_e}{m_i} \frac{v_A^2}{v_{t,i}^2} (\frac{r_{L,A,i}}{L})^2
\end{aligned}$$

2. Start from (4.27) and find the system of equations (4.28). Rewrite (4.28) as a linear system of equations in its standard form for the unknowns j_x, j_y, j_z and obtain the solution (4.29), (4.30), (4.31).

3. Work out the limiting cases $\omega_{c,e} = 0$, $\omega_{c,e} << \overline{\nu}_{e,i}$ and $\omega_{c,e} >> \overline{\nu}_{e,i}$ for $\sigma_{\|}$, $\sigma_{\perp}$ and $\sigma_{\wedge}$. (BT1997, W1986).

4. Determine the graphs of $\sigma_{\|}$, $\sigma_{\perp}$ and $\sigma_{\wedge}$ as functions of $\omega_{c,e}/\overline{\nu}_{e,i}$ in the interval $[0, 10]$. (BT1997).

5. Denote by $\tilde{\eta}$ the tensor inverse of σ so that $\sigma.\tilde{\eta} = \mathbf{1}$. Show that

$$\tilde{\eta} = \begin{bmatrix} \tilde{\eta}_{\perp} & -\tilde{\eta}_{\wedge} & 0 \\ \tilde{\eta}_{\wedge} & \tilde{\eta}_{\perp} & 0 \\ 0 & 0 & \tilde{\eta}_{\|} \end{bmatrix}$$

where

$$\tilde{\eta}_{\|} = \frac{1}{\sigma_{\|}}, \quad \tilde{\eta}_{\perp} = \frac{\sigma_{\perp}}{\sigma_{\perp}^2 + \sigma_{\wedge}^2}, \quad \tilde{\eta}_{\wedge} = \frac{-\sigma_{\wedge}}{\sigma_{\perp}^2 + \sigma_{\wedge}^2}$$

(W1986).

6. Derive the induction equation in ideal and resistive MHD

$$\begin{aligned}
\frac{\partial \vec{B}}{\partial t} &= \nabla \times \left(\vec{v} \times \vec{B}\right) \quad \text{ideal MHD} \\
&= \nabla \times \left(\vec{v} \times \vec{B}\right) + \eta \nabla^2 \vec{B} \quad \text{resistive MHD}
\end{aligned}$$

7. In this problem you try to understand under what circumstances we can neglect thermal conduction. This you do by comparing the thermal conduction term to the ideal term

$$\nabla.(\kappa_0 T^{5/2} \nabla T) \quad \rightarrow \quad \frac{\rho^\gamma}{\gamma - 1} \frac{d}{dt} \frac{p}{\rho^\gamma}$$

by a scale analysis. Take values in the solar corona

$$T = 10^6 \text{K}, \quad p = 0.01 \text{ Pa}, \quad \tau = 1\text{h}$$

and compute the corresponding length that your magnetic field line should have for thermal conduction to be unimportant! (H2000)

8. Rewrite the induction equation of ideal MHD as

$$\frac{d\vec{B}}{dt} = -\vec{B}(\nabla.\vec{v}) + (\vec{B}.\nabla)\vec{v}.$$

$\frac{d\vec{B}}{dt}$ is the variation in time of the magnetic field when we follow the motion of the plasma element. Interpret the two terms in the right hand side of this equation.

9. Determine the dimension of η! (H2000)

10. Compute R_m and the diffusion time τ_η for a sunspot! Use

$$L = 10^7 m,\ \ \eta = 10^3 \mathrm{m^2 s^{-1}},\ \ v = 10^3 \mathrm{ms^{-1}}.$$

(H2000)

11. Compute the plasma β for a coronal active region where the magnetic field is closed and for a coronal hole. In an active region

$$\begin{aligned} B &= 100 \text{ Gauss} = 10^{-2}\text{Tesla},\ \ \mu = 4\pi \times 10^{-7}\text{H m}^{-1},\ \ \overline{\mu} = 0.5,\ \ n = 10^{16}\text{m}^{-3}, \\ m_P &= 1.6726 \times 10^{-27}\text{kg},\ \ T = 2 \times 10^6\text{K} \end{aligned}$$

Coronal holes have a weaker magnetic field strength and a lower temperature. Take

$$B = 10 \text{ Gauss} = 10^{-3}\text{Tesla},\ \ \rho = 1.67 \times 10^{-12}\text{kg m}^{-3},\ \ T = 1 \times 10^6\text{K}$$

(H2000)

12. Consider the diffusion of a current sheet. Start from an initial anti-parallel magnetic field

$$\begin{aligned} \vec{B} &= B_0(x)\vec{1}_z \\ &= B_0\vec{1}_z\ \ x > 0 \\ &= -B_0\vec{1}_z\ \ x < 0 \end{aligned}$$

Compute how this initial discontinuity diffuses away at $x = 0$ assuming that the field remains in the $z-$direction for all time. The equation that you have to solve is

$$\frac{\partial B}{\partial t} = \eta \frac{\partial^2 B}{\partial^2 x}$$

subject to the initial condition $B = B_0(x)$. (H2000)

13. Consider the kinematic concentration of the magnetic field due to the granular motion in the solar photosphere. We can model granulation by an array of cells with a flow

given by the velocity

$$\vec{v} = (v_x, 0, v_z)^t$$

$$v_x = v_0 \sin(kx)$$

$$v_z = -k\, z\, v_0\, \sin(kx)$$

To make analytical progress neglect the back reaction of the magnetic field on the flow and only consider the induction equation. (H2000).

14. Fill in the gaps in the derivation of Ferraro's isorotation law.

15. Look in Cartesian coordinates (x, y, z) at the following magnetic fields

$$\vec{B} = B_0 \vec{1}_x,$$

$$\vec{B} = B_0 e^y \vec{1}_x,$$

$$\vec{B} = B_0(\vec{1}_x - 2x\vec{1}_y)$$

$$\vec{B} = B_0(\vec{1}_x + 2x\vec{1}_y)$$

$$\vec{B} = B_0(2y\vec{1}_x - \vec{1}_y)$$

$$\vec{B} = B_0(2y\vec{1}_x + \vec{1}_y)$$

$$\vec{B} = B_0(y\vec{1}_x + \alpha^2 x\vec{1}_y), \quad \alpha^2 > 1$$

Determine for each of these magnetic fields the equation of the field lines and sketch the field lines for 5 equidistant values of the integration constant in the equation of the field lines. Calculate the electric current, the Lorentz force, the magnetic pressure, the magnetic pressure force and the magnetic tension force. (P1985, H2000).

16. Look in cylindrical coordinates (r, φ, z) at the magnetic field $\vec{B}(r) = B(r)\vec{1}_\varphi$ with

$$B(r) = \begin{cases} r & \text{if } r \leq 1 \\ 1/r & \text{if } r > 1 \end{cases}$$

Sketch the magnetic field lines. Calculate the Lorentz force, the magnetic pressure force and the magnetic tension force in $r \leq 1$ and in $r > 1$. Are your results consistent with your intuitive expectations about pressure.

Joseph Larmor (1857 - 1942)

He went on to become Lucasian Professor of Mathematics at Cambridge in 1903, the chair becoming vacant on the death of Stokes in February of that year.
He published three papers all entitled A dynamical theory of the electric and luminiferous medium between 1894 and 1897. These papers presented his theory of the electron, which of course gained further weight in 1897 when J. J. Thomson experimentally identified the electron.
Larmor wrote Aether and Matter in 1900 (renamed by Lamb Aether and no matter) which was a winning entry for the Adams Prize at Cambridge in 1898. It incorporated much of the work of the three major papers of 1894-1897 we referred to above. Warwick writes in [14]:- His book of 1900, Aether and matter, Cambridge University Press, Cambridge, 1900, helped to establish a research school that guided the development of mathematical electromagnetic theory in Cambridge until the end of World War I. However Warwick [14] also writes:- Today, however, Larmor is widely remembered by scientists for just two formulae and one theorem which, although correctly attributed to him, have been seen by historians of science as tangential to his main research interests. Indeed, none of the recent scholarly studies of Larmor's scientific work even mention the now famous formulae and theorem.
We should take Warwick's lead and make sure that we mention those concepts to which Larmor's name is attached today. These are the 'Larmor precession', the 'Larmor frequency', 'Larmor's theorem' and 'Larmor's formula'. The first explains the splitting and polarisation of the spectral lines in a magnetic field. The Larmor frequency relates to electrons orbiting in a magnetic field and led him to postulate electrons as orbiting around some centre. He appears to have been the first to predict this behaviour. Larmor's theorem is a related result concerning how a certain transformation can negate the magnetic field for a charged particle subject to electric and magnetic fields. He was the first to calculate the rate of energy radiation from an accelerating electron and for this he gave Larmor's formula which gives the power radiated in terms of the electron's charge and acceleration. The formula breaks down for velocities close to the speed of light due to relativistic effects. When George Stokes and William Thomson (Lord Kelvin) died, Larmor acted as an editor for their complete works. He also brought out a new version of Henry Cavendish's works in 1921, Maxwell had been the editor for the original publication.
Larmor retired from the Lucasian Chair of Mathematics at Cambridge in 1932. He was succeeded in this position by Dirac.
Larmor was conservative in temperament, questioning modern trends even in such matters as the installation of baths in the College (1920). "We have done without them for 400 years, why begin now?", he once said at a College meeting. Yet once the innovation were made he was a regular user. Morning by morning in a mackintosh and cap, in which he was not seen at other times, he found his way across the bridge to the New Court baths.
http://www-history.mcs.st-andrews.ac.uk/history/Mathematicians/
Article by: J. J. O'Connor and E. F. Robertson

Chapter 5

Basic MHD dynamics

> People my age are hung up on the Sixties. Everybody knows that and regards it as sort of a problem with us: the generation who won't throw out their bell-bottoms. ⋯
> So naturally I think something special happened in the sixties. Didn't it? Or was it just because I was that age, between things, when everything was still possible, that time, which in retrospect, doesn't seem to last long?
> *Michael Frain*
> "The survivor's Guide", September 4, 1992.
> *The Laws of our Fathers*
> Scott Turow

This Chapter is concerned with the dynamics of magnetized plasmas in the very-low-frequency large-scale approximation of Magnetohydrodynamics. The focus is on the basic MHD waves that can occur in a magnetized plasma. We forget about the gravitational force and are concerned with the MHD waves that are driven by the plasma pressure force and the Lorentz force. This leads us to Alfvén waves and magnetosonic waves. The study of these waves enable us to identify the Alfvén velocity, v_A, and the velocity of sound, v_S, as fundamental quantities characterizing the very-low-frequency large-scale fluid-like behaviour of plasmas. Magneto-gravity waves will be absent for the present discussion. There are situations in solar and astrophysical plasmas where a gravitational force, either due to an external gravitational field or due to self-gravitation, has to be included in the mathematical analysis.

5.1 Linear motions superimposed on a static equilibrium

The previous Section ended with the prediction of transversal waves on magnetic field lines which are driven by the magnetic tension force; these are the famous Alfvén waves. Alfvén waves are not the only waves that can exist in a magnetized plasma. Since a plasma is compressive we can anticipate acoustic waves possibly modified by the magnetic field. MHD waves, and also MHD instabilities are solutions of the equations of (ideal) MHD. If you have a good look at the set of ideal MHD equations you come to realize that we are dealing with a complicated set of non-linear partial differential equations (PDE). This set of PDE's have to be supplemented with boundary conditions if the plasma that we are looking at occupies a finite volume (which in reality it always does!). Let there be no doubt in your mind that the set of MHD PDE's is horrendously complicated from a mathematical and computational point of view because it contains a lot of physics.

Fortunately, nature gives us a helping hand. Often, the amplitudes of the wave variables are small compared to their equilibrium values. This fortunate situation enables us to linearize the set of original non-linear MHD equations. The linearized versions of the MHD equations are much more tractable to mathematical analysis and numerical simulations. Linear differential equations are well documented in mathematics. Linear theory of MHD waves is a valid approximation of reality if the plasma configuration under study occurs in a state of (static or stationary) equilibrium during a time span that is sufficiently longer than the dynamical time scale of the system, which is the Alfvén transit time. In thermonuclear fusion the aim is to create magnetic plasmas that exist for a sufficiently long time to realize fusion of hydrogen ions. So we are definitely interested in equilibrium configurations and in the MHD waves and MHD instabilities that can occur in these fusion plasmas. In the solar atmosphere we see magnetic structures that exist during days, weeks and months. So, nature provides us with occasions where it is of obvious interest to know about equilibrium states.

Ideal MHD is the adequate framework for studying three types of problems. Firstly, ideal MHD provides us with the mathematical tool for computing realistic static (or stationary) equilibrium configurations. The mathematical formulation of equilibrium configurations involves non-linear partial differential equations. Secondly, this equilibrium model has to be tested on its dynamic stability with respect to small perturbations. This can be done with the use of linear ideal MHD. An equilibrium model is physically meaningful only if it is dynamically stable or if it is at least dynamically stable with respect to perturbations with short growth rates. Thirdly, we want to know what kind of waves can occur in this stable magnetic equilibrium configuration. An obvious reason for our interest in MHD waves is that they are carriers of energy. MHD waves can transfer energy from one part of the plasma to another part. If there is a way for dissipating part of the energy carried by the MHD waves you have a mechanism for heating plasma. Note that dissipation requires going beyond ideal MHD and including dissipative effects in the MHD equations! Linear ideal MHD is a natural starting point for the study of MHD waves.

MHD stability and MHD waves are two complementary facets of the dynamics of a plasma. MHD stability is concerned with the unstable motions of the plasma that terminate the equilibrium while MHD waves are the stable motions that can persist in the plasma equilibrium for long time spans until they are damped by dissipation. It is no surprise that they both can be studied with *linear ideal* MHD. It is good to be aware of the limitations of linear ideal MHD. Because it is *linear*, linear ideal MHD cannot give you any information on the non-linear evolution of a linearly unstable perturbation and it cannot give you any information on the stability with respect to large amplitude motions. Because it is *ideal*, linear ideal MHD cannot give you any information on dissipative instabilities. Because it is *linear*, the use of linear ideal MHD is restricted to *small amplitude* MHD waves. Because it is *ideal*, linear ideal MHD cannot give you any information on the possible damping or excitation of linear ideal MHD waves. In spite of this (incomplete) list of limitations, linear ideal MHD can give us a lot of information on the dynamics of magnetic plasmas.

The aim is to study linear motions superimposed on a static equilibrium or background. Our starting point is the ideal MHD version of the set of non-linear partial differential equations (4.35). For our present purposes the Eulerian version with partial time derivatives is to be preferred. The equations that govern the linear motions superimposed on a static equilibrium state are obtained by linearizing the original MHD equations around this static equilibrium.

A plasma is in equilibrium if it does not change in time

$$\frac{\partial}{\partial t} = 0$$

and it is in static equilibrium if in addition

$$\vec{v} = 0$$

Under these conditions the equations of ideal MHD (4.35) can be simplified to the equations for a magnetostatic equilibrium

$$\boxed{\begin{aligned} -\nabla p + \rho \vec{g} + \frac{1}{\mu}(\nabla \times \vec{B}) \times \vec{B} &= 0 \\ \nabla . \vec{B} &= 0 \quad (5.1) \end{aligned}}$$

The first equation expresses a balance between the mechanical forces and the Lorentz force (note that we shall drop the gravitational force in what follows). An exact MHD equilibrium state does not exist in the real world. There are always temporal changes in a MHD system. They are due to MHD waves (and MHD instabilities) and to the diffusion of the magnetic field due to electrical resistivity. Recall that the diffusion of the magnetic field happens on the electromagnetic diffusion time scale τ_η. Hence, the notion of a static equilibrium only makes sense on time spans much shorter than τ_η. The time scale of the MHD waves in a magnetic plasma is the Alfvén crossing τ_A and the notion of a static equilibrium only makes sense on time spans much longer than τ_A. Hence

$$\frac{L}{v_A} = \tau_A \ll \tau_{MSE} \ll \tau_\eta = Lu \frac{L}{v_A}$$

with τ_{MSE} the time scale of the magnetostatic equilibrium. An obvious necessary condition is that $Lu \gg 1$. Let us now go back to the original non-linear MHD equations and obtain the equations that govern the linear motions around a static equilibrium state. We need to make a distinction between the distributions of the physical variables, density, pressure, magnetic field and velocity, in the equilibrium state and in the perturbed state where motions are superimposed on the equilibrium state. The distributions of the physical variables in the equilibrium state are denoted by a subscript 0. Hence

$$\rho_0(\vec{r}), \;\; p_0(\vec{r}), \;\; \vec{B}_0(\vec{r}), \;\; \vec{v}_0(\vec{r}) \equiv 0$$

are the equilibrium distributions. They satisfy the equations of magnetostatic equilibrium (5.1). Let us now superimpose motions on the equilibrium state so that the plasma element that is at the position $\vec{r}$ is displaced to the position $\vec{r} + \vec{\xi}$

$$\vec{r} \quad \rightarrow \quad \vec{r}+\vec{\xi} \qquad (5.2)$$

$\vec{\xi}$ is called the *Lagrangian displacement*. The motions cause changes in density, pressure and magnetic field. We can measure these changes in a fixed geometrical position (Eulerian description) or while following the motion of the plasma (Lagrangian description). If we denote the physical quantities in the static equilibrium as $f_0(\vec{r})$ and in the time dependent plasma as $f(\vec{r};t)$ then

$$f'(\vec{r};t) = f(\vec{r};t) - f_0(\vec{r}) \qquad (5.3)$$

is the change in the quantity f in a fixed geometrical position. It is the *Eulerian* perturbation of f. On the other hand

$$\delta f(\vec{r};t) = f(\vec{r}+\vec{\xi};t) - f_0(\vec{r}) \qquad (5.4)$$

gives the change in f while we follow the motion of the fluid. It is the *Lagrangian* perturbation of f. In linear theory the Lagrangian and Eulerian variation of f are simply related as

$$\delta f(\vec{r};t) = f'(\vec{r};t) + \vec{\xi}.\nabla f_0(\vec{r})$$

The equations that govern the linear motions superimposed on a static equilibrium state are obtained by linearizing the ideal MHD equations. After this straightforward process we find

$$\begin{aligned}
\rho' &= -\nabla.(\rho_0\vec{\xi}) \\
p' &= -\vec{\xi}.\nabla p_0 - \gamma p_0 \nabla.\vec{\xi} \\
\vec{B}' &= \nabla \times \left(\vec{\xi} \times \vec{B_0}\right) \\
\rho_0 \frac{\partial^2 \vec{\xi}}{\partial t^2} &= -\nabla p' + \frac{1}{\mu}(\nabla \times \vec{B}_0) \times \vec{B}' + \frac{1}{\mu}(\nabla \times \vec{B}') \times \vec{B}_0 \qquad (5.5)
\end{aligned}$$

The right hand side of the linearized version of the equation of motion can be written in terms of $\vec{\xi}$, since both p' and $\vec{B}'$ can be written in terms of $\vec{\xi}$. Hence, we can write the linearized version of the equation of motion with the aid of the force operator $\vec{F}(\vec{\xi})$ as

$$\rho_0 \frac{\partial^2 \vec{\xi}}{\partial t^2} = \vec{F}(\vec{\xi})$$

where

$$\vec{F}(\vec{\xi}) = \nabla\left(\vec{\xi}.\nabla p_0 + \gamma p_0 \nabla.\vec{\xi}\right) + \frac{1}{\mu}(\nabla \times \vec{B}_0) \times \vec{B}' + \frac{1}{\mu}(\nabla \times \vec{B}') \times \vec{B}_0$$

In order to make it very clear that $\vec{B}'$ is also written in terms of $\vec{\xi}$ we sometimes use the notation

$$\vec{B}' = \vec{Q}(\vec{\xi}) = \nabla \times \left(\vec{\xi} \times \vec{B_0}\right)$$

This equation shows that there are only three unknown scalar functions in linear ideal MHD, namely the three components of the Lagrangian displacement $\vec{\xi}(\vec{r};t)$. This is in sharp contrast to non-linear ideal MHD with its eight unknown scalar functions, the three components of velocity $\vec{v}$, the three components of $\vec{B}$, density and pressure.

The equations for the linear motions of an incompressible plasma can be obtained by taking the limit

$$v_S^2 \to \infty \text{ or } \gamma \to \infty \text{ combined with } \nabla.\vec{\xi} = 0$$

so that the Eulerian and Lagrangian perturbations of pressure are finite

$$\delta p = \rho_0 v_S^2 \nabla.\vec{\xi} \text{ and } p' \text{ are finite.}$$

v_S^2 is the square of the local speed of sound and is defined in (4.14). The equations for the linear motions of an incompressible plasma are

$$\begin{aligned} \nabla.\vec{\xi} &= 0 \\ \vec{B}' &= \nabla \times \left(\vec{\xi} \times \vec{B_0}\right) \\ \rho_0 \frac{\partial^2 \vec{\xi}}{\partial t^2} &= -\nabla p' + \frac{1}{\mu}(\nabla \times \vec{B_0}) \times \vec{B}' + \frac{1}{\mu}(\nabla \times \vec{B}') \times \vec{B_0} \end{aligned} \quad (5.6)$$

p' cannot be expressed in terms of $\vec{\xi}$. Hence, we have four unknowns functions, the three components of $\vec{\xi}$ and p'. The problem is well defined since $\nabla.\vec{\xi} = 0$ combined with the linearized equation of motion provide us with four scalar equations. The reader might be confused by the limit $v_S^2 \to \infty$ in a Chapter that is using MHD to study very-low-frequency waves in magnetized non-relativistic plasmas. What is meant by this limit is that v_S is much larger than any other fluid-like velocity in the plasma. In the present situation the only other fluid-like velocity is the Alfvén velocity, v_A.

The equilibrium states are time independent by definition. As a consequence the linearized version of the equation of motion seen as a differential equation in time has constant coefficients. This means that we can look for solutions of the form

$$f(\vec{r};t) = \tilde{f}(\vec{r}) \exp(-i\sigma t) \quad (5.7)$$

where $\tilde{f}(\vec{r})$ is the time independent part of the solution and f stands for any of the perturbed quantities. If we leave the system to its own devices, meaning that we do not give the system a kick at some moment in time or we do not impose a perturbation during some time interval on a specific part of the system, then we are looking at the normal modes of oscillation of the system. These normal modes are governed by the equation

$$-\sigma^2 \rho_0 \vec{\xi}(\vec{r}) = \vec{F}(\vec{\xi}(\vec{r}))$$

or

$$\sigma^2 \vec{\xi}(\vec{r}) = \vec{L}(\vec{\xi}(\vec{r})), \quad \vec{L}(\vec{\xi}(\vec{r})) = -\frac{1}{\rho_0}\vec{F}(\vec{\xi}(\vec{r}))$$

$\vec{L}$ and $\vec{F}$ are linear and time independent operators. The eigenvalues of the operator $\vec{L}$ are the squares of the eigenfrequencies of the plasma system that we are looking at and the corresponding eigenvectors are the eigenmodes. It can be shown that the operator $\vec{L}$ in combination with the appropriate boundary conditions is Hermitian with respect to the scalar product

$$(\vec{\xi}_1, \vec{\xi}_2) = \int_V \rho_0 \vec{\xi}_1 . \vec{\xi}_2^* dV$$

This result is important since it implies that the eigenvalues σ^2 are real and that the eigenvectors are mutually orthogonal. The fact that the eigenvalues σ^2 are real means that we have two types of solutions:

- For $\sigma^2 > 0$ the solutions vary harmonically in time as $\cos(\sigma t)$. These stable solutions correspond to oscillations and waves.
- For $\sigma^2 < 0$ the solutions vary exponentially as $\exp(-\gamma t)$ and as $\exp(\gamma t)$. Here $\gamma^2 = -\sigma^2$, $\gamma > 0$ and both solutions are present. The exponentially decaying solution is of little importance. The exponentially increasing solution corresponds to a dynamical instability of the plasma equilibrium. Linear theory only applies to the initial linear stage of a dynamical instability.

Complex values for σ^2 can occur in ideal MHD for so-called quasi-modes and for MHD waves that are allowed to leak energy out of the system. We shall not be concerned here with these subtleties.

5.2 Waves of a uniform plasma of infinite extent

Let us now try to understand the MHD waves of the most simple plasma equilibrium: a uniform plasma of infinite extent. Of course plasmas are not uniform and they do not have infinite extensions. Nevertheless, a uniform plasma of infinite extent is a good place to start our discovery journey in the wonderland of MHD waves. For a uniform plasma of infinite extent the MHD waves exist in their most pure form. We shall be able to put the animals in separate cages. The results for a uniform plasma of infinite extent will help us understand the MHD waves in more complicated structured and non-uniform plasmas. If we do not understand the MHD waves in the simplest situation, we cannot hope to understand them in more complicated situations.

In (5.7) we have already factored out the time dependence by putting the perturbed quantities proportional to $\exp(-i\sigma t)$. Since the background is uniform, the coefficients in the system of the partial differential equations are constants. Consequently we can look for solutions in the form of plane waves and write the spatial part $\tilde{f}(\vec{r})$ of any wave variable f as:

$$\tilde{f}(\vec{r}) = \hat{f} \exp(i\vec{k}.\vec{r}) = \hat{f} \exp(i(k_x x + k_y y + k_z z))$$

Combining the temporal and spatial factors we find

$$\begin{aligned} f(\vec{r}; t) &= \tilde{f}(\vec{r}) \exp(-i\sigma t) = \hat{f} \exp(i(\vec{k}.\vec{r} - \sigma t)) \\ &= \hat{f} \exp(i(k_x x + k_y y + k_z z - \sigma t)) \end{aligned} \qquad (5.8)$$

$\hat{f}$ is the constant amplitude of f, $\vec{k} = k_x\vec{1}_x + k_y\vec{1}_y + k_z\vec{1}_z$ is the wave vector and σ is the frequency of the wave. Since $\partial/\partial t$ and ∇ only act on $\exp(i(\vec{k}.\vec{r} - \sigma t))$ it follows that

$$\frac{\partial}{\partial t} \rightarrow -i\sigma, \quad \nabla \rightarrow i\vec{k}$$

For a uniform plasma (5.5) and (5.6) can be simplified by noting that

$$\nabla p_0 = 0, \quad \nabla \rho_0 = 0, \quad \nabla \times \vec{B}_0 = 0, \quad \nabla.\vec{B}_0 = 0$$

$$\begin{aligned}
\rho' &= -i\rho_0(\vec{k}.\vec{\xi}) \\
p' &= -i\rho_0 v_S^2(\vec{k}.\vec{\xi}) \\
\vec{B}' &= i\vec{k} \times \left(\vec{\xi} \times \vec{B}_0\right) = i\vec{\xi}(\vec{k}.\vec{B}_0) - i\vec{B}_0(\vec{k}.\vec{\xi}) \\
-\rho_0\sigma^2\vec{\xi} &= -i\vec{k}p' + \frac{i}{\mu}(\vec{k} \times \vec{B}') \times \vec{B}_0 \\
&= -\rho_0 v_S^2\vec{k}(\vec{k}.\vec{\xi}) - i\vec{k}\frac{\vec{B}_0.\vec{B}'}{\mu} + i\frac{\vec{k}.\vec{B}_0}{\mu}\vec{B}' \qquad (5.9)
\end{aligned}$$

for a compressible plasma; and

$$\begin{aligned}
\vec{k}.\vec{\xi} &= 0 \\
\rho' &= 0 \\
\vec{B}' &= i\vec{k} \times \left(\vec{\xi} \times \vec{B}_0\right) = i\vec{\xi}(\vec{k}.\vec{B}_0) \\
-\rho_0\sigma^2\vec{\xi} &= -i\vec{k}p' + \frac{i}{\mu}(\vec{k} \times \vec{B}') \times \vec{B}_0 \\
&= -i\vec{k}p' + \vec{k}\frac{\vec{k}.\vec{B}_0}{\mu}(\vec{B}_0.\vec{\xi}) - \frac{(\vec{k}.\vec{B}_0)^2}{\mu}\vec{\xi} \qquad (5.10)
\end{aligned}$$

for an incompressible plasma.

5.3 Sound waves

Instead of trying to solve the set of equations (5.9) in its full form, we take a gradual approach and start with the simplest possible situation and subsequently add more complexity to the analysis. All of us are familiar with this most simple situation as we all know about acoustic waves. We do not need a magnetic field to have acoustic waves. Hence let us look at a compressible non-magnetic plasma. When we insert the expression for p' in the equation of motion of our set of equations (5.9) we find

$$\rho_0 \sigma^2 \vec{\xi} = \rho_0 \, v_S^2 \, \vec{k} \, (\vec{k}.\vec{\xi})$$

This equation tells us that the displacement vector $\vec{\xi}$ and the wave vector $\vec{k}$ are parallel. The obvious quantity to use in this situation is

$$Y = \vec{k}.\vec{\xi} \qquad (5.11)$$

which is the compression since

$$\nabla.\vec{\xi} = i\vec{k}.\vec{\xi} = iY$$

The equation for Y is readily obtained as

$$\rho_0(\sigma^2 - k^2 v_S^2)Y = 0 \qquad (5.12)$$

Non-trivial solutions with $Y \neq 0$ exist only for

$$\sigma^2 = k^2 v_S^2 \qquad (5.13)$$

This is the acoustic wave. It is a completely isotropic and compressive wave which is driven by the plasma pressure. Its phase velocity $\vec{v}_{ph}$ and group velocity $\vec{v}_{gr}$ are equal

$$\vec{v}_{ph} = \frac{\sigma}{k^2}\vec{k} = v_S \vec{1}_k, \quad \vec{v}_{gr} = \nabla_{\vec{k}}\sigma = v_S \vec{1}_k$$

The group velocity is the velocity with which the energy in the wave is transported. The velocity of sound v_S is identified here as the velocity at which the phase and energy propagate in a sound wave in a plasma in absence of a magnetic field. Hence, it is a fundamental quantity related to the very-low-frequency large-scale fluid-like behaviour of plasmas.

5.4 Alfvén waves

From the previous Section we know that plasma pressure and compressibility in absence of any magnetic field generate sound waves. In this Section we try to understand what kind of waves are generated by the Lorentz force. In particular we are interested in the magnetic tension force. For that reason we subject the magnetic field lines to displacements perpendicular to the magnetic field lines. As we do not want any interference from plasma pressure we restrict our attention to incompressible motions. So, it should be very clear to the reader that we are trying to find out what happens when Ferraro's law of isorotation is violated.

The displacement is incompressible and perpendicular to the magnetic field lines

$$\vec{k}.\vec{\xi} = 0, \quad \vec{B}_0.\vec{\xi} = 0$$

The displacement vector $\vec{\xi}$ is perpendicular to both $\vec{k}$ and $\vec{B}_0$. When $\vec{k}$ and $\vec{B}_0$ are not parallel, then $\vec{\xi}$ is parallel to $\vec{B}_0 \times \vec{k}$ and an obvious quantity to consider here is

$$Z = (\vec{1}_B \times \vec{k}).\vec{\xi} = (\vec{k} \times \vec{\xi}).\vec{1}_B \qquad (5.14)$$

which is the component of the rotation of the displacement along the direction of the equilibrium field since

$$(\nabla \times \vec{\xi}).\vec{1}_B = i(\vec{k} \times \vec{\xi}).\vec{1}_B$$

The analysis for propagation along the magnetic field lines ($\vec{k} \parallel \vec{B}_0$) is not different from that for oblique propagation. The displacement $\vec{\xi}$ is not confined to a line but to a plane normal to $\vec{B}_0$. There is no magnetic pressure force and the Lorentz force is reduced to the magnetic tension force

$$LF = \vec{T}'_n = -\frac{(\vec{k}.\vec{B}_0)^2}{\mu}\vec{\xi}$$

The equation of motion of the set of equations (5.10) can be written as

$$\rho_0 \left\{ \sigma^2 - \frac{(\vec{k}.\vec{B}_0)^2}{\mu\rho_0} \right\} \vec{\xi} = i\vec{k}p'$$

Since $\vec{k}.\vec{\xi} = 0$, it follows that

$$\rho_0 \left\{ \sigma^2 - \frac{(\vec{k}.\vec{B}_0)^2}{\mu\rho_0} \right\} \vec{\xi} = 0, \;\; p' = 0 \qquad (5.15)$$

Since $\vec{\xi}$ is parallel to $\vec{B}_0 \times \vec{k}$, (5.15) can be seen as an equation for Z. A non-trivial solution with $\vec{\xi} \neq 0$ exists if and only if

$$\sigma^2 = \sigma_A^2 \qquad (5.16)$$

σ_A is the Alfvén frequency and is defined as

$$\sigma_A^2 = \frac{(\vec{k}.\vec{B}_0)^2}{\mu\rho_0} = \frac{k_{||}^2 B_0^2}{\mu\rho_0} = k_{||}^2 v_A^2 = k^2 v_A^2 \cos^2\theta \qquad (5.17)$$

$k_{||}$ is the component of the wave vector $\vec{k}$ parallel to the equilibrium magnetic field $\vec{B}_0$ and θ is the angle between the equilibrium magnetic field $\vec{B}_0$ and the wave vector $\vec{k}$. This solution is the classic Alfvén wave. The Alfvén wave is a purely magnetic wave driven solely by magnetic tension, which we now can write as (see Problem 5)

$$\vec{T}'_n = -\rho_0\sigma_A^2\vec{\xi}$$

There are not any variations in density and plasma pressure: $\rho' = 0, \;\; p' = 0$. The displacement of the Alfvén wave is perpendicular to the equilibrium magnetic field ($\vec{B}_0.\vec{\xi} = 0$) and to the wave vector ($\vec{k}.\vec{\xi} = 0$).

The Alfvén wave is highly anisotropic. Its frequency depends only on the component of the wave vector parallel to the magnetic field. This implies that the group velocity is always directed along the equilibrium magnetic field. The Alfvén wave transports energy solely along the magnetic field lines and in no other direction. The eigenvalue $\sigma^2 = \sigma_A^2 = k_{||}^2 v_A^2$ is (infinitely) degenerate since it only depends on one component of the wave vector, i.e. $k_{||}$ and is independent of the two remaining components.

The equilibrium configuration is a uniform plasma of infinite extent. In this equilibrium configuration there is only one preferred direction and that is the direction defined by the equilibrium magnetic field $\vec{B}_0$. We can exploit this fact by adopting a system of field aligned Cartesian coordinates (x, y, z) in which the wave vector $\vec{k}$ plays a special role. We choose the x-axis along the constant equilibrium magnetic field, so that

$$\vec{B}_0 = B_0 \vec{1}_x$$

For a given wave vector $\vec{k}$ the constant equilibrium magnetic field $\vec{B}_0$ and the wave vector $\vec{k}$ define a plane. We choose the z-axis in that plane so that

$$\vec{k} = k_x \vec{1}_x + k_z \vec{1}_z, \quad k_y = 0$$

Since for the Alfvén wave the displacement is perpendicular to both the equilibrium field and the wave vector $\vec{k}$, it follows that

$$\boxed{\vec{\xi}_A = \xi_y \vec{1}_y \qquad (5.18)}$$

With this choice of field aligned coordinates

$$\sigma_A^2 = k_x^2 v_A^2 = k^2 v_A^2 \cos^2\theta$$

so that

$$\vec{v}_{ph,A} = \frac{\sigma_A}{k^2}\vec{k} = v_A \cos\theta \vec{1}_k, \quad \vec{v}_{gr,A} = \nabla_{\vec{k}}\sigma_A = v_A \vec{1}_x \qquad (5.19)$$

The phase velocity changes in magnitude from v_A for propagation parallel to $\vec{B}_0$ to 0 for propagation perpendicular to $\vec{B}_0$. The group velocity is always oriented along the magnetic field and in magnitude equal to v_A. Hence, v_A emerges here as a fundamental quantity related to the very-low-frequency large-scale fluid-like behaviour of plasmas. Since there are not any forces other than the plasma pressure force and the Lorentz force present in the MHD description of plasmas, v_S and v_A are the two fundamental velocities associated with the very-low-frequency large-scale fluid-like behaviour of plasmas. All the other velocities that will show up in the discussion of ideal MHD waves are combinations of v_S and v_A. The degeneracy of the eigenvalue $\sigma^2 = \sigma_A^2 = k_x^2 v_A^2$ is obvious here since the eigenvalue is independent of k_z.

The Alvén wave is highly anisotropic. The phase velocity and in particular the group velocity illustrate this extreme anisotropy. The variations of the phase velocity and of the group velocity with the angle θ can be easily illustrated with the help of the polar diagrams of these quantities. The polar diagram of a 2-dimensional vector field that depends on 1 variable, $\vec{p}(\theta)$, is the parametric planar curve defined by the equations $x = p_x(\theta), z = p_z(\theta)$. The polar diagram of the phase velocity for Alfvén waves consists of two circles with their centres at $(v_A/2, 0)$ and $(-v_A/2, 0)$ respectively and with radius $v_A/2$. The polar diagram of the group velocity for Alfvén waves consists of the two points $(v_A, 0)$ and $(-v_A, 0)$. The reader is asked to draw his/her polar diagrams for Alfvén waves in Problem 6. Incidentally, the polar diagram of the phase and group velocity of the sound waves that we found in the previous Section are a circle with its centre at the origin and radius v_S.

5.5 Alfvén waves and slow waves

In the previous Section we have restricted the motions to being incompressible and perpendicular to the equilibrium magnetic field. The outcome was the Alfvén wave. In the present Section we drop the second restriction and allow the displacement vector to have components parallel and normal to the equilibrium magnetic field. The displacement is incompressible $\vec{k}.\vec{\xi} = 0$, but not necessarily perpendicular to $\vec{B}_0$ so that $\vec{B}_0.\vec{\xi} \neq 0$. The Lorentz force is

$$LF = \vec{k}\frac{\vec{k}.\vec{B}_0}{\mu}(\vec{B}_0.\vec{\xi}) - \rho_0\sigma_A^2\vec{\xi}$$

The first term is the magnetic pressure force and the second term is the magnetic tension force. The equation of motion of the set of equations (5.10) can be written as

$$\rho_0(\sigma^2 - \sigma_A^2)\vec{\xi} = i\vec{k}\left\{p' + i\frac{\vec{k}.\vec{B}_0}{\mu}(\vec{B}_0.\vec{\xi})\right\}$$

For parallel propagation along the magnetic field lines ($\vec{k} \parallel \vec{B}_0$) we have $\vec{B}_0.\vec{\xi} = 0$ in addition to $\vec{k}.\vec{\xi} = 0$ and we are back in the previous Section on Alfvén waves. When $\vec{k}$ and $\vec{B}_0$ are not parallel, then these two vectors combined with $\vec{B}_0 \times \vec{k}$ are three linearly independent vectors. They can be transformed into three orthogonal vectors

$$\vec{1}_B,\ \ \vec{k} - (\vec{k}.\vec{1}_B)\vec{1}_B,\ \ \vec{1}_B \times \vec{k}$$

It is then convenient to use the components of $\vec{\xi}$ along these orthogonal directions

$$X = \vec{1}_B.\vec{\xi} = \xi_\parallel,\ \ W = (\vec{k} - (\vec{k}.\vec{1}_B)\vec{1}_B).\vec{\xi},\ \ Z = (\vec{1}_B \times \vec{k}).\vec{\xi} = (\vec{k} \times \vec{\xi}).\vec{1}_B \qquad (5.20)$$

W can be expressed in terms of X by using the constraint $\vec{k}.\vec{\xi} = 0$ so that $W = -k_\parallel\xi_\parallel$. Since $\vec{k}.\vec{\xi} = 0$, it follows that

$$\rho_0(\sigma^2 - \sigma_A^2)X = 0,\ \ \rho_0(\sigma^2 - \sigma_A^2)Z = 0,\ \ p' + i\frac{\vec{k}.\vec{B}_0}{\mu}B_0X = 0 \qquad (5.21)$$

There are two types of solutions

- $\sigma^2 = \sigma_A^2,\ \ Z \neq 0,\ \ X = 0,\ \ W = 0,\ \ \rho' = 0,\ \ p' = 0$
- $\sigma^2 = \sigma_A^2,\ \ Z = 0,\ \ X \neq 0,\ \ W = -k_\parallel X,\ \ \rho' = 0,\ \ p' = -i\frac{\vec{k}.\vec{B}_0}{\mu}B_0X$

The first solution corresponds to waves that have no displacement component along the magnetic field $X = \vec{1}_B.\vec{\xi} = 0$. The displacement is also perpendicular to the wave vector since $\vec{k}.\vec{\xi} = 0$. In addition there are no variations in density and pressure. These are the Alfvén waves.

The second solution corresponds to a wave which has a displacement component along the magnetic field. There is no variation in density. Plasma pressure and magnetic pressure vary, although in anti-phase so that total pressure is constant. This is the slow magnetosonic wave which in an incompressible plasma has the same frequency as the Alfvén wave.

Let us now, as in the previous Section, adopt a system of field aligned Cartesian coordinates (x, y, z). It is straightforward to show that

$$\vec{1}_B = \vec{1}_x, \ \vec{k} - (\vec{k}.\vec{1}_B)\vec{1}_B = k_z\vec{1}_z, \ \vec{1}_B \times \vec{k} = -k_z\vec{1}_y$$

$$X = \xi_x, \ W = k_z\xi_z, \ Z = -k_z\xi_y$$

The constraint $\vec{k}.\vec{\xi} = 0$ implies that $\xi_z = -(k_x/k_z)\xi_x$.
The equations (5.21) reduce to the simpler form

$$\rho_0(\sigma^2 - \sigma_A^2)\xi_x = 0, \ \rho_0(\sigma^2 - \sigma_A^2)\xi_y = 0, \ p' + i\rho v_A^2 k_x\xi_x = 0$$

In this system the Alfvén wave and the slow magnetosonic wave are characterized by

- $\sigma^2 = \sigma_A^2, \ \xi_y \neq 0, \ \xi_x = 0, \ \xi_z = 0, \ \rho' = 0, \ p' = 0$
- $\sigma^2 = \sigma_A^2, \ \xi_y = 0, \ \xi_x \neq 0, \xi_z = -k_x\xi_x/k_z, \ p' = -i\rho v_A^2 k_x\xi_x$

Hence,

$$\boxed{\vec{\xi}_A = \xi_y\vec{1}_y, \ \vec{\xi}_{sl} = \xi_x(\vec{1}_x - \frac{k_x}{k_z}\vec{1}_z) \qquad (5.22)}$$

The displacement of the Alfvén wave is perpendicular to both the equilibrium magnetic field and the wave vector. The displacement of the slow wave is confined to the plane defined by the wave vector and the constant equilibrium magnetic field.

Here also the eigenvalue $\sigma^2 = \sigma_A^2 = k_x^2 v_A^2$ is (infinitely) degenerate in the sense it is independent of k_z. In addition to the infinite degeneracy with respect to k_z there is a double degeneracy since we have two (fundamentally different) solutions (respectively normal to the plane defined by $\vec{k}$ and by $\vec{B}_0$ and in that plane) for a given set (k_x, k_z).

The phase velocity and the group velocity are the same as those found for the Alfvén wave in the previous Section.

5.6 Alfvén waves and magnetosonic waves

Let us now drop the restriction that the motions are incompressible. We go back to the equation of motion and the induction equation of (5.9). We compute the Lorentz force and find (see Problem 10) that

$$LF = -\rho_0\sigma_A^2\vec{\xi} + \frac{(\vec{k}.\vec{B}_0)}{\mu}(\vec{\xi}.\vec{B}_0)\vec{k} - \frac{B_0^2}{\mu}(\vec{k}.\vec{\xi})\vec{k} + \frac{(\vec{k}.\vec{B}_0)}{\mu}(\vec{k}.\vec{\xi})\vec{B}_0$$

We now choose a system of Cartesian coordinates with the $x-$axis parallel to the constant equilibrium magnetic field so that

$$\vec{B}_0 = B_0\vec{1}_x, \ \vec{k}.\vec{B}_0 = k_x B_0, \ \vec{\xi}.\vec{B}_0 = B_0\xi_x$$

In this system of coordinates we can write the equation of motion (5.9) as

$$\sigma^2\vec{\xi} = v_S^2(\vec{k}.\vec{\xi})\vec{k} + \sigma_A^2\vec{\xi} - k_x v_A^2\xi_x\vec{k} + v_A^2(\vec{k}.\vec{\xi})\vec{k} - k_x v_A^2(\vec{k}.\vec{\xi})\vec{1}_x$$

It is straightforward to write down the three components of this vector equation (see Problem 11). These three equations are a set of three linear and homogeneous equations for the unknowns ξ_x, ξ_y, ξ_z that we can write using matrix notation as

$$\sigma^2 \vec{\xi} = M \vec{\xi}$$

with M a real and symmetric 3×3 matrix

$$M = \begin{bmatrix} k_x^2 v_S^2 & k_x k_y v_S^2 & k_x k_z v_S^2 \\ k_x k_y v_S^2 & k_y^2 (v_S^2 + v_A^2) + \sigma_A^2 & k_y k_z (v_S^2 + v_A^2) \\ k_x k_z v_S^2 & k_y k_z (v_S^2 + v_A^2) & k_z^2 (v_S^2 + v_A^2) + \sigma_A^2 \end{bmatrix}$$

The wave modes are the eigenvectors and the squares of the frequencies are the eigenvalues of the matrix M. Since the matrix M is real and symmetric its eigenvalues are real and the eigenvectors that belong to distinct eigenvalues are orthogonal. The eigenvalues (i.e. the squares of the frequencies σ^2) are the solutions of the characteristic equation

$$\det(M - \sigma^2 I_3) = 0$$

As you know from your course on Linear Algebra the sum of the eigenvalues is equal to the sum of the diagonal elements of the matrix M and the product of the eigenvalues is equal to the determinant of M :

$$\sum_{l=1}^{3} \sigma_l^2 = k^2 (v_S^2 + v_A^2) + k_x^2 v_A^2, \quad \prod_{l=1}^{3} \sigma_l^2 = \det M$$

In spite of this information the actual calculation of the three eigenvalues of matrix M is rather complicated. Basically mathematics is telling us that we have not been clever in choosing the physical quantities for describing MHD waves in a plasma. ξ_x, ξ_y, ξ_z are not the best variables for characterizing the MHD waves. From the previous three Sections we know variables that give a better characterization. The sound waves are characterized by compression measured by Y. The Alfvén waves are characterized by rotation around the equilibrium magnetic field; this is measured by Z. Finally the slow magnetosonic waves need a displacement component along the equilibrium magnetic field, which is ξ_x. Even if we had not taken the time to go through the successive steps of the previous three Sections we could have identified this set of variables tailored to the study of MHD waves by the following argumentation. The (constant) magnetic field specifies a preferred direction in the plasma. In our system of Cartesian coordinates this is the $x-$direction, so the component in this preferred direction, ξ_x is a natural variable. Next an important property of motion is whether or not it causes compressions and expansions of the plasma. This is measured by $\nabla.\vec{\xi}$, which for our plane wave solutions reduces to $i\vec{k}.\vec{\xi}$. Our second natural variable is $Y = \vec{k}.\vec{\xi}$. Finally, it is important to know whether the motion causes the plasma to rotate. Since there is only one preferred direction in the present situation, the direction of $\vec{B}$ or the $x-$direction, we are interested in rotation around this preferred direction. This is measured by $(\nabla \times \vec{\xi}).\vec{1}_x$, which for our plane wave solutions reduces to $i(\vec{k} \times \vec{\xi}).\vec{1}_x$. Our third natural variable is $Z = (\vec{k} \times \vec{\xi}).\vec{1}_x$. The three natural variables for describing the MHD waves are

$$\xi_x, \quad Y = \vec{k}.\vec{\xi}, \quad Z = (\vec{k} \times \vec{\xi}).\vec{1}_x = k_y \xi_z - k_z \xi_y \tag{5.23}$$

With these variables the components of the equation of motion can be replaced with a far simpler set of equations (see Problem 12)

$$\begin{aligned} \sigma^2 \xi_x - k_x v_S^2 Y &= 0 \\ k^2 v_A^2 k_x \xi_x + [\sigma^2 - k^2(v_S^2 + v_A^2)]Y &= 0 \\ (\sigma^2 - \sigma_A^2) Z &= 0 \end{aligned} \tag{5.24}$$

It is decoupled into two subsets. The first subset consists of the first two equations which contain the unknowns ξ_x and Y. The third equation only contains the unknown Z.

Eigenfrequencies and eigenvectors

Hence, the solutions are as follows.

The Alfvén wave

$$\xi_x = 0, \quad Y = 0, \quad Z \neq 0, \quad \sigma^2 = \sigma_A^2 = k_{||}^2 v_A^2 \tag{5.25}$$

This solution is the classic Alfvén wave. It is a transversal wave with its displacement perpendicular to the magnetic field lines. Since $Y = 0$, the Alfvén wave does not cause any compression or expansion of the plasma and density and plasma pressure remain unchanged: $\rho' = 0, \ p' = 0$. It is a purely magnetic wave driven by the magnetic tension force.

The magnetosonic waves

$$\xi_x \neq 0, \quad Y \neq 0, \quad Z = 0 \tag{5.26}$$

These solutions are associated with the eigenvalues σ^2 that are the roots of the characteristic equation

$$\det \begin{bmatrix} \sigma^2 & -k_x v_S^2 \\ k_x k^2 v_A^2 & \sigma^2 - k^2(v_S^2 + v_A^2) \end{bmatrix} = 0$$

or

$$(\sigma^2)^2 - k^2(v_S^2 + v_A^2)\sigma^2 + k_x^2 k^2 v_S^2 v_A^2 = 0 \tag{5.27}$$

This is the eigenvalue equation for the (slow and fast) magnetosonic waves. The solutions to this equation are

$$\sigma_{sl,f}^2 = \frac{k^2(v_S^2 + v_A^2)}{2} \left\{ 1 \pm \left(1 - \frac{4\sigma_C^2}{k^2(v_S^2 + v_A^2)} \right)^{1/2} \right\} \tag{5.28}$$

σ_C is the cusp frequency. It is defined as

$$\sigma_C^2 = \frac{v_S^2}{v_S^2 + v_A^2}\sigma_A^2 = k_{||}^2 v_C^2 = k_x^2 v_C^2, \quad v_C^2 = \frac{v_S^2 v_A^2}{v_S^2 + v_A^2} \tag{5.29}$$

v_C is called the cusp velocity. The index "sl" (slow) corresponds to the minus sign and the index "f" (fast) corresponds to the plus sign. It is straightforward to show that the frequencies σ_{sl}, σ_A and σ_f satisfy the following sequences of inequalities (see Problem 14)

$$\sigma_C^2 \le \sigma_{sl}^2 \le \sigma_A^2 \le k^2 v_A^2 \le \sigma_f^2$$

$$\sigma_C^2 \le \sigma_{sl}^2 \le k_x^2 v_S^2 \le k^2 c^2 \le \sigma_f^2$$

$$\sigma_C^2 \le \sigma_{sl}^2 \le \min(k_x^2 v_S^2, k_x^2 v_A^2)$$

$$\max(k^2 v_S^2, k^2 v_A^2) \le \sigma_f^2 \tag{5.30}$$

The inequalities (5.30) are used to call the waves associated with σ_{sl} the *slow magnetosonic waves* and those associated with σ_f the *fast magnetosonic waves*.

Let us have a look at the following limiting cases (see Problem 15)

- No magnetic field $B = 0$
  Obviously $v_A^2 = 0 \;\wedge\; v_C^2 = 0$ so that

  $$\sigma_{sl}^2 = 0, \quad \sigma_A^2 = 0, \quad \sigma_f^2 = k^2 v_S^2$$

  No slow waves, no Alfvén waves, just the classic acoustic waves driven by gas pressure!

- Weak magnetic field $v_A^2 \ll v_S^2$
  It is straightforward to show that

  $$\sigma_{sl}^2 \approx k_x^2 v_A^2 = \sigma_A^2, \quad \sigma_f^2 \approx k^2 v_S^2$$

  The fast wave is the classic acoustic wave modified by the magnetic field. The slow wave is now a magnetic wave which is weakly affected by acoustic effects. Although its frequency only slightly differs from that of the Alfvén wave, the slow wave and the Alfvén wave are very different waves (why?).

- Incompressible plasma $v_S^2 \to +\infty$
  It is easy to show that $\lim_{v_S^2 \to +\infty} v_C^2 = v_A^2$, $\lim_{v_S^2 \to +\infty} \sigma_C^2 = \sigma_A^2$ and hence

  $$\sigma_{sl}^2 = \sigma_A^2, \quad \sigma_f^2 = +\infty$$

  The fast wave has infinite frequency. This is no surprise since the assumption of an incompressible plasma means that acoustic signals travel with infinite speed and consequently have infinite frequency (period is zero!). The slow wave is a magnetic wave. Its frequency is equal to the Alfvén frequency, but the slow wave and the Alfvén wave are basically different.

- Strong magnetic field $v_S^2 \ll v_A^2$
  It is easy to show that

$$\sigma_{sl}^2 \approx k_x^2 v_S^2 \quad \sigma_f^2 \approx k^2 v_A^2$$

  The slow wave is an acoustic wave modified by the magnetic field. It is definitely different from the classic acoustic wave (why?). The fast wave is now a magnetic wave which is mainly driven by magnetic pressure and weakly affected by acoustic effects.

- No plasma pressure $v_S^2 = 0$
  Obviously $v_C^2 = 0, \quad \sigma_C^2 = 0$ and

$$\sigma_{sl}^2 = 0, \quad \sigma_f^2 = k^2 v_A^2$$

  The slow wave has disappeared. The fast wave is a magnetic wave solely driven by magnetic pressure.

We can summarize these results as

	slow wave	fast wave
$\vec{B} = 0$	$\sigma_{sl}^2 = 0$	$\sigma_f^2 = k^2 v_S^2$
$v_A^2 \ll v_S^2$	$\sigma_{sl}^2 \approx \sigma_A^2$	$\sigma_f^2 \approx k^2 v_S^2$
$v_S^2 \to +\infty$	$\sigma_{sl}^2 = \sigma_A^2$	$\sigma_f^2 = +\infty$
$v_S^2 \ll v_A^2$	$\sigma_{sl}^2 \approx k_x^2 v_S^2$	$\sigma_f^2 \approx k^2 v_A^2$
$v_S^2 = 0$	$\sigma_{sl}^2 = 0$	$\sigma_f^2 = k^2 v_A^2$

(5.31)

The slow wave is always anisotropic, while the fast wave is isotropic. The slow wave is a magnetic wave for weak magnetic fields and an acoustic wave for strong magnetic fields. The situation is reversed for the fast wave. It is an acoustic wave for weak magnetic fields and a magnetic wave for strong magnetic fields.

The eigenvectors

Let us now have a look at the spatial eigenvectors. A straightforward calculation yields the following results (see Problem 16)

$$\rho' = -i\rho_0 Y$$

$$p' = -i\rho_0 v_S^2 Y$$

$$\vec{B}' = ik_x B_0 \vec{\xi} - i\vec{B}_0 Y$$

$$\begin{aligned} p'_m &= \frac{\vec{B}'.\vec{B}_0}{\mu} = ik_x\rho_0 v_A^2\xi_x - i\rho_0 v_A^2 Y \\ p'_t &= ik_x\rho_0 v_A^2\xi_x - i\rho_0(v_S^2 + v_A^2)Y \\ \frac{1}{\mu}(\vec{B}_0.\nabla)\vec{B}' &= ik_x p'_m \vec{1}_x - \rho_0\sigma_A^2(\xi_y\vec{1}_y + \xi_z\vec{1}_z) \\ \vec{T}'_n &= -ik_x p'_m\vec{1}_x + \frac{1}{\mu}(\vec{B}_0.\nabla)\vec{B}' = -\rho_0\sigma_A^2(\xi_y\vec{1}_y + \xi_z\vec{1}_z) \end{aligned}$$

The reader has already guessed that p'_m and p'_t are the Eulerian perturbation of the magnetic pressure and of total pressure respectively. Let us adopt our system of field aligned Cartesian coordinates (x, y, z) once again so that

$$Y = k_x\xi_x + k_z\xi_z, \quad Z = -k_z\xi_y$$

The Alfvén wave

From (5.25) we know that $\xi_x = 0,$, $Y = 0,$, $Z \neq 0$ so that

$$\begin{aligned} \vec{\xi}_A &= \xi_y\vec{1}_y, \quad \vec{B}' = ik_x B_0\xi_y\vec{1}_y, \quad \vec{T}'_n = -\rho_0\sigma_A^2\xi_y\vec{1}_y \\ \rho' &= 0, \quad p' = 0, \quad p'_m = 0, \quad p'_t = 0 \end{aligned} \qquad (5.32)$$

The magnetosonic waves

From (5.26) we know that $\xi_x \neq 0$, $Y \neq 0$, $Z = 0$. In addition ξ_x and Y are related by the first two equations of (5.24). These two equations are equivalent for the eigenfrequencies of the magnetosonic waves, i.e. the eigenfrequencies that satisfy the eigenvalue equation (5.27). We can rewrite this eigenvalue relation as (see Problem 17)

$$\frac{k_z^2 v_S^2}{\sigma_{sl,f}^2 - k_x^2 v_S^2} = \frac{\sigma_{sl,f}^2 - k^2 v_A^2}{\sigma_{sl,f}^2}$$

and find that (see Problem 18)

$$\xi_x = \frac{k_x v_S^2}{\sigma_{sl,f}^2} Y = \frac{k_z^2 v_S^2}{\sigma_{sl,f}^2 - k_x^2 v_S^2}\frac{k_x}{k_z}\xi_z = \frac{\sigma_{sl,f}^2 - k^2 v_A^2}{\sigma_{sl,f}^2}\frac{k_x}{k_z}\xi_z$$

Hence

$$\vec{\xi}_{sl,f} = \xi_x(\vec{1}_x + \frac{\sigma_{sl,f}^2}{\sigma_{sl,f}^2 - k^2 v_A^2}\frac{k_z}{k_x}\vec{1}_z) = \xi_z(\frac{\sigma_{sl,f}^2 - k^2 v_A^2}{\sigma_{sl,f}^2}\frac{k_x}{k_z}\vec{1}_x + \vec{1}_z) \qquad (5.33)$$

It is straightforward to show that $\vec{\xi}_{sl}$ and $\vec{\xi}_f$ are orthogonal. It is obvious that they are also orthogonal to $\vec{\xi}_A$, so that

$$\vec{\xi}_A, \quad \vec{\xi}_{sl}, \quad \vec{\xi}_f$$

are three orthogonal eigenvectors that span the 3-dimensional space of the displacement vectors (see Problem 18). Let us have a look at the other physical quantities, compression Y, magnetic field, density, plasma pressure, magnetic pressure and total pressure. We can rewrite the relation between ξ_x and ξ_z as

$$k_x \xi_x = \frac{\sigma_{sl,f}^2 - k^2 v_A^2}{\sigma_{sl,f}^2} \frac{k_x^2}{k_z^2} k_z \xi_z$$

Since

$$\frac{\sigma_f^2 - k^2 v_A^2}{\sigma_f^2} > 0, \quad \frac{\sigma_{sl}^2 - k^2 v_A^2}{\sigma_{sl}^2} < 0$$

it follows that $k_x \xi_x$ and $k_z \xi_z$ have the same sign for fast waves and opposite signs for slow waves so that for a displacement vector of given magnitude and for given wave vector, the compression (in absolute value) is always largest for the fast waves.

A straightforward calculation yields the following results for the magnetosonic waves (see Problem 18)

$$\begin{aligned}
\vec{B}' &= iB_0 \xi_z (-k_z \vec{1}_x + k_x \vec{1}_z) \\
p' &= -i\rho_0 v_S^2 Y \\
p'_m &= i\rho_0 v_A^2 \frac{k_x^2 v_S^2 - \sigma_{sl,f}^2}{\sigma_{sl,f}^2} Y \\
\frac{p'_m}{p'} &= \frac{v_A^2}{v_S^2} \frac{\sigma_{sl,f}^2 - k_x^2 v_S^2}{\sigma_{sl,f}^2} = \frac{k_z^2 v_A^2}{\sigma_{sl,f}^2 - k^2 v_A^2} \\
\frac{p'_t}{p'} &= \frac{v_S^2 + v_A^2}{v_S^2} \frac{\sigma_{sl,f}^2 - k_x^2 v_C^2}{\sigma_{sl,f}^2} = \frac{\sigma_{sl,f}^2 - k_x^2 v_A^2}{\sigma_{sl,f}^2 - k^2 v_A^2} \\
\vec{T}'_n &= -\rho_0 \sigma_A^2 \xi_z \vec{1}_z
\end{aligned} \tag{5.34}$$

Since

$$\sigma_f^2 - k_x^2 v_S^2 > 0, \quad \sigma_f^2 - k^2 v_A^2 > 0$$

$$\sigma_{sl}^2 - k_x^2 v_S^2 < 0, \quad \sigma_{sl}^2 - k^2 v_A^2 < 0$$

it follows that the plasma pressure and magnetic pressure are in phase for the fast waves ($p'_m/p' > 0$) and half a period out of phase for the slow waves. In other words, for the fast waves plasma pressure and magnetic pressure work together while they are working against each other for the slow waves. This behaviour is most prominent for propagation perpendicular to the magnetic field lines. Indeed

$$\lim_{k_x \to 0} \left(\frac{p'_m}{p'}\right)_{sl} = \lim_{k_x \to 0} \frac{k_z^2 v_A^2}{\sigma_{sl}^2 - k^2 v_A^2} = -1$$

so that

$$(p_t')_{sl} = 0$$

for slow waves with a propagation vector perpendicular to the magnetic field lines. On the other hand the maximal value of the ratio p_t'/p' for fast waves occurs for perpendicular propagation $k_x = 0$

$$\max(\frac{p_t'}{p'})_f = \frac{v_S^2 + v_A^2}{v_S^2}$$

Phase velocities

The Alfvén waves and the slow magnetosonic waves are very anisotropic. Let us have a look at the phase velocities. Recall that

$$\vec{v}_{ph} = \frac{\sigma}{k^2}\vec{k}$$

Alfvén waves

The phase velocity of the Alfvén waves is well known by now. According to (5.19) $\vec{v}_{ph,A} = v_A \cos\theta \vec{1}_k$ with θ the angle between the magnetic field and the propagation vector in the $xz-$plane. The dependence of the (magnitude of the) phase velocity on the direction of propagation, i.e. the angle θ is contained in the factor $\cos\theta$. The phase velocity is maximal for $\theta = 0$, i.e. for propagation along the magnetic field lines and minimal and equal to zero for $\theta = \pi/2$ i.e. for perpendicular propagation. Alfvén waves do not propagate perpendicular to the magnetic field lines and propagate fastest along the field lines.

Magnetosonic waves

The phase velocity of the magnetosonic waves follows from (5.28). It is

$$\vec{v}_{ph} = \vec{1}_k \frac{(v_S^2 + v_A^2)^{1/2}}{\sqrt{2}} \left\{ 1 \pm \left(1 - \frac{4v_C^2 \cos^2\theta}{(v_S^2 + v_A^2)} \right)^{1/2} \right\}^{1/2} \tag{5.35}$$

The dependence of the phase velocity (5.35) on the angle θ is now more complicated than for the Alfvén waves. Two interesting limiting cases are parallel and perpendicular propagation. For parallel propagation

$$v_{ph,sl} = \min(v_S,\ v_A),\quad v_{ph,f} = \max(v_S,\ v_A)$$

For perpendicular propagation

$$v_{ph,sl} = 0,\quad v_{ph,f} = (v_S^2 + v_A^2)^{1/2}$$

The phase velocity for fast waves is minimal for parallel propagation and maximal for perpendicular propagation. It varies from $\max(v_S, v_A)$ to $(v_S^2 + v_A^2)^{1/2}$ so that the variation is never larger than 41% (when does this situation occurs?). So, all in all, the fast waves are rather isotropic. The phase velocity of the slow waves is minimal for perpendicular propagation and maximal for parallel propagation. It varies from 0 to $\min(v_S, v_A)$. Slow waves are definitely anisotropic; they refuse to propagate perpendicular to the magnetic field lines. In that respect they are like Alfvén waves. A polar diagram of the phase velocity for the slow and fast waves is shown on Fig. 5.1. The reader is asked to draw his/her own polar diagrams for slow and fast magnetosonic waves in Problem 20.

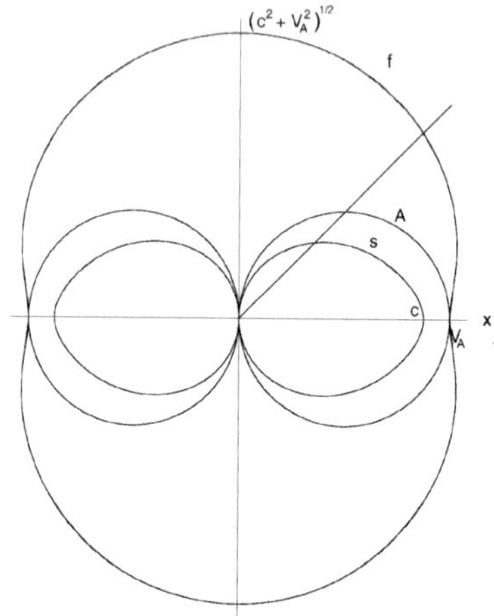

Figure 5.1: *Polar diagram of phase velocity for slow and fast magnetosonic waves and Alfvén waves,* $v_A > v_S$.

Eigenfrequencies again: dependence on direction of propagation

For Alfvén waves the dependence of the eigenfrequency on the wave vector is very simple. In a system of Cartesian coordinates with the x−axis aligned with the equilibrium magnetic field, the (square of the) frequency of the Alfvén wave is (5.25) $\sigma_A^2 = k_x^2 v_A^2$. It is independent of k_z (infinitely degenerate) and varies as k_x^2.

For magnetosonic waves the dependence of the frequency on the wave vector is more complicated than that for Alfvén waves. The frequencies of the fast and slow waves are given by (5.28):

$$\sigma_{sl,f}^2 = \frac{k^2(v_S^2 + v_A^2)}{2} \left\{ 1 \pm \left(1 - \frac{4\sigma_C^2}{k^2(v_S^2 + v_A^2)} \right)^{1/2} \right\}$$

- σ^2 as a function of k_x for fixed k_z.
  For $k_x = 0$ we have
  $$\sigma_{sl}^2 = 0, \quad \sigma_f^2 = k_z^2(v_S^2 + v_A^2)$$
  For $\lim_{k_x \to \infty}$ we have (see problem 20)
  $$\sigma_{sl}^2 \approx \min(k_x^2 v_A^2, k_x^2 v_S^2), \quad \sigma_f^2 \approx \max(k_x^2 v_A^2, k_x^2 v_S^2)$$
  The variation of $\sigma_A^2, \sigma_{sl}^2, \sigma_f^2$ as function of k_x for fixed k_z is shown on Fig. 5.2. The reader computes and subsequently draws his/her figure showing $\sigma_A^2, \sigma_{sl}^2, \sigma_f^2$ as function of k_x for fixed k_z (see Problem 20).

- σ^2 as a function of k_z for fixed k_x.
  For $k_z = 0$ we have
  $$\sigma_{sl}^2 = \sigma_I^2, \quad \sigma_f^2 = \sigma_{II}^2$$

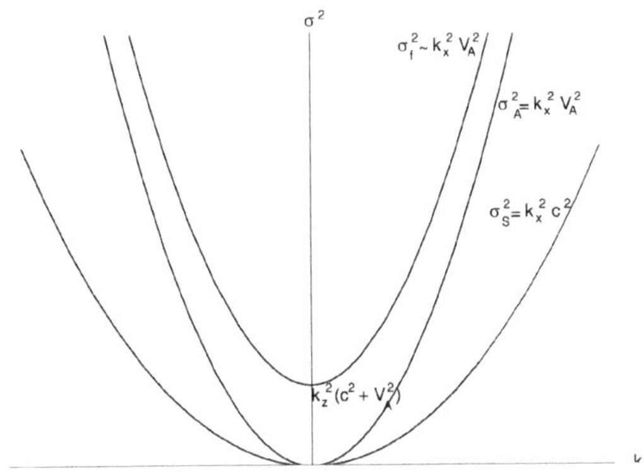

Figure 5.2: *Variation of σ^2 as function of k_x for fixed k_z ($k_y = 0$), $v_A > v_S$*

where σ_I, σ_{II} are the so-called cut-off frequencies, these are the frequencies at which the slow respectively fast waves emerge in the spectrum when σ is viewed as a function of k_z. These cut-off frequencies are defined as

$$\sigma_I^2 = \min(k_x^2 v_S^2, k_x^2 v_A^2), \quad \sigma_{II}^2 = \max(k_x^2 v_S^2, k_x^2 v_A^2) \tag{5.36}$$

For $\lim_{k_z \to \infty}$ we have (see problem 21)

$$\lim_{k_z \to \infty} \sigma_{sl}^2 = k_x^2 v_C^2 = \sigma_C^2, \quad \sigma_f^2 \approx k_z^2 (v_S^2 + v_A^2)$$

Hence the cusp frequency σ_C is the accumulation point of the frequencies of the slow magnetosonic waves. The frequencies of the fast magnetosonic waves have an accumulation point at ∞.

We can now rewrite the last two lines of (5.30) as

$$\underbrace{\sigma_C^2 \leq \sigma_{sl}^2 \leq \sigma_I^2}_{\text{slow}} \leq \sigma_A^2 \leq \underbrace{\sigma_{II}^2 \leq \sigma_f^2 \to \infty}_{\text{fast}} \tag{5.37}$$

which shows that

$$\sigma_C^2, \ \sigma_I^2, \ \sigma_A^2, \ \sigma_{II}^2 \tag{5.38}$$

are the four fundamental characteristic frequencies that give a natural division of the spectrum into the slow subspectrum, the Alfvén subspectrum and the fast subspectrum. The variation of $\sigma_A^2, \sigma_{sl}^2, \sigma_f^2$ as function of k_z for fixed k_x is shown on Fig. 5.3.

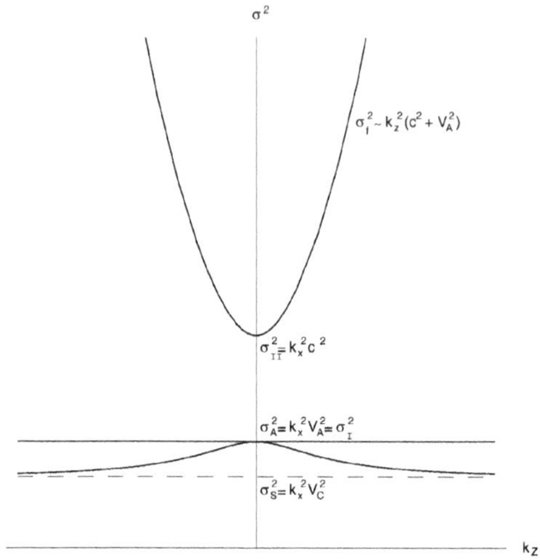

Figure 5.3: *Variation of σ^2 as function of k_z for fixed k_x ($k_y = 0$), $v_S > v_A$*

Group velocities

The group velocity is the velocity with which waves transport their energy. It is given by

$$\vec{v}_{gr} = \nabla_{\vec{k}}\sigma$$

Alfvén waves

The group velocity of the Alfvén waves is well-known by now. According to (5.19) it is $\vec{v}_{gr,A} = v_A \vec{1}_x$. Alfvén waves transport their energy exclusively along the magnetic field lines and do it with the velocity of Alfvén v_A. This is anisotropic behaviour in its most extreme form.

Magnetosonic waves

As before we take a field aligned system of Cartesian coordinates so that $\vec{k}$ is in the $xz-$plane. With that choice

$$\vec{v}_{gr} = [\frac{\partial\sigma}{\partial k_x}, 0, \frac{\partial\sigma}{\partial k_z}]^t$$

We use the solutions to the dispersion relation for magnetosonic waves (5.28) and find by straightforward calculation (see Problem 22)

$$v_{gr,x} = v_{ph}\cos\theta \mp \frac{v_C^2}{v_{ph}}\frac{\cos\theta\sin^2\theta}{(1-\Delta)^{1/2}}, \quad v_{gr,z} = v_{ph}\sin\theta \pm \frac{v_C^2}{v_{ph}}\frac{\cos^2\theta\sin\theta}{(1-\Delta)^{1/2}}$$

where

$$\Delta(\theta) = 4\frac{v_C^2}{v_A^2 + v_S^2}\cos^2\theta$$

These expressions are indeterminate for the slow wave when $\theta \to \pi/2$ since $\lim_{\theta\to\pi/2} v_{ph,sl} = 0$. We can lift this indeterminacy by rewriting the expressions for $v_{gr,x}$ and $v_{gr,z}$ as

$$v_{gr,x} \quad = \quad v_{ph}\cos\theta \mp \frac{v_C}{\sqrt{2}}\frac{[1 \mp (1-\Delta)^{1/2}]^{1/2}}{(1-\Delta)^{1/2}}\sin^2\theta$$

$$v_{gr,z} = v_{ph} \sin\theta \pm \frac{v_C}{\sqrt{2}} \frac{[1 \mp (1-\Delta)^{1/2}]^{1/2}}{(1-\Delta)^{1/2}} \sin\theta\cos\theta \tag{5.39}$$

The dependence of the group velocity (5.39) on the angle θ is far more complicated than for the Alfvén waves. Two interesting limiting cases are parallel and perpendicular propagation. For parallel propagation ($\theta = 0,\ \vec{k} = k_x \vec{1}_x$) we find

$$\vec{v}_{gr,sl} = \min(v_S, v_A)\vec{1}_x,\ \ \vec{v}_{gr,f} = \max(v_S, v_A)\vec{1}_x$$

Here the group velocity of both the slow and fast wave are in the direction of the wave vector and equal to the corresponding phase velocity.

For perpendicular propagation ($\theta = \pi/2,\ \vec{k} = k_z \vec{1}_z$) we find

$$\vec{v}_{gr,sl} = v_C \vec{1}_x,\ \ \vec{v}_{gr,f} = (v_S^2 + v_A^2)^{1/2}\vec{1}_z$$

The group velocity of fast wave is in the direction of the wave vector and equal to the phase velocity of the fast wave. For the slow wave we have the remarkable result that the group velocity is along the magnetic field although the propagation is perpendicular to the magnetic field.

For a general propagation angle θ the group velocity and phase velocity make an angle. The group and phase velocities have the same direction for parallel and perpendicular propagation for fast waves and for parallel propagation for slow waves. The deviation between group velocity and phase velocity is small for the fast waves, but rather large for slow waves. Also it is straightforward to show that (see Problem 22)

$$v_{gr,z} k_z > 0 \text{ for fast waves}$$

$$v_{gr,z} k_z < 0 \text{ for slow waves.}$$

The result for fast waves is what you expect. Phase and energy are both propagated upward or both downward. For slow waves we have the strange result that a wave with an upward propagating phase propagates its energy downwards and vice versa. The polar diagrams of the group velocity of the slow and fast waves are shown in Fig. 5.4. The magnitude of the group velocity of the fast wave is minimal for parallel propagation and maximal for perpendicular propagation. It varies from $\max(c, v_A)$ to $(v_S^2 + v_A^2)^{1/2}$. This is the same range as for the phase velocity of the fast wave. The direction of the group velocity only slightly deviates from the direction of propagation. Hence, the fast waves are isotropic. The group velocity of the slow wave is confined to a small cone around the magnetic field line. Slow waves transport energy only in a small cone around the magnetic field lines. The slow wave is highly anisotropic, it is almost as anisotropic as the Alfvén wave!

The reader is asked to draw his/her own polar diagrams for the group velocities for slow and fast magnetosonic waves in Problem 24.

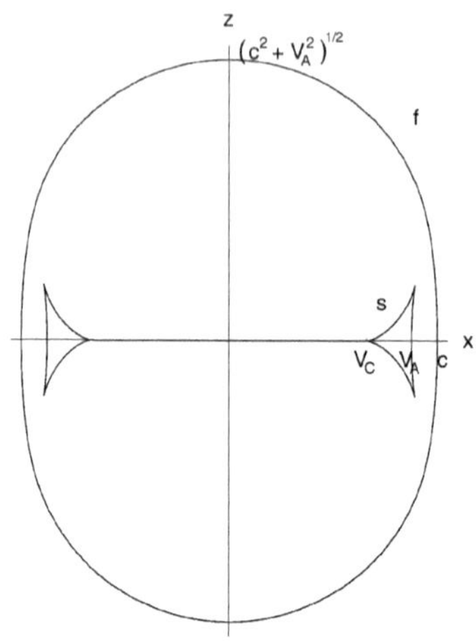

Figure 5.4: *Group velocity of slow and fast magnetosonic waves and of Alfv'en waves* ($k_y = 0$), $v_S > v_A$.

Summary

1. Alfvén waves

   - They are driven by the magnetic tension force.
   - They have no density and pressure variations.
   - The displacements are perpendicular to the wave vector and to the equilibrium magnetic field.
   - They are anisotropic and cannot propagate perpendicular to the magnetic field lines.
   - The energy flows along the magnetic field lines at the Alfvén velocity.

2. Fast magnetosonic waves

   - They are driven by tension and pressure forces.
   - They have density and pressure variations.
   - Plasma pressure and magnetic pressure variations are in phase.
   - The displacements are confined to the plane defined by the wave vector and the equilibrium magnetic field and have parallel and perpendicular components with respect to the wave vector and to the equilibrium magnetic field.
   - The flow of energy is fairly isotropic.
   - They are pretty isotropic with a slight preference to directions perpendicular to the equilibrium magnetic field.

3. Slow magnetosonic waves

   - They are driven by tension and pressure forces.
   - They have density and pressure variations.
   - Plasma pressure and magnetic pressure variations are in anti-phase.
   - The displacements are confined to the plane defined by the wave vector and the equilibrium magnetic field and have parallel and perpendicular components with respect to the wave vector and to the equilibrium magnetic field.
   - The flow of energy is highly anisotropic and confined to a small cone around the magnetic field lines.
   - They are highly anisotropic and cannot propagate in directions perpendicular to the equilibrium magnetic field.

A first step towards understanding MHD waves in non-uniform plasmas

In the previous subsection we have often used a system of field aligned coordinates with the $x-$axis along the magnetic field and the $xz-$plane defined by the equilibrium magnetic field and the wave vector $\vec{k}$. The latter we could do as we had only one preferred direction in our system, the direction of the magnetic field. This is not always possible or, more correctly, is seldom possible. When we move from a uniform plasma of infinite extent to a non-uniform plasma of finite extent this will be no longer possible. In addition, to the preferred direction defined by the magnetic field, we will have at least one preferred direction defined by the variation of the equilibrium quantities of the static background. So as a first step toward understanding waves in a non-uniform background we use a wave vector $\vec{k} = k_x\vec{1}_x + k_y\vec{1}_y + k_z\vec{1}_z$ with $k_y \neq 0$. If we want to anticipate effects related to the finite extent of the background plasma we can look at quantization prescriptions for e.g. k_z.

The frequencies of the Alfvén waves and of the the slow and fast magnetosonic waves are given by (5.16), (5.17) and (5.28). The squares of these frequencies are

$$\sigma_A^2 = k_x^2 v_A^2, \quad \sigma_{sl,f}^2 = \frac{k^2(v_S^2 + v_A^2)}{2}\left\{1 \pm \left(1 - \frac{4k_x^2 v_C^2}{k^2(v_S^2 + v_A^2)}\right)^{1/2}\right\}$$

with $k^2 = k_x^2 + k_y^2 + k_z^2$. When we look at the frequencies as functions of k_z we can define the cut-off frequencies σ_I and σ_{II} as the frequencies at which the slow and fast waves emerge in the spectrum. Put $k_z = 0$ in the expressions for $\sigma_{sl,f}^2$ and find σ_I and σ_{II} as

$$\begin{aligned}
\sigma_{I,II}^2 &= \frac{1}{2}(k_x^2 + k_y^2)(v_S^2 + v_A^2)\left\{1 \pm \left(1 - \frac{4k_x^2 v_C^2}{(k_x^2 + k_y^2)(v_S^2 + v_A^2)}\right)^{1/2}\right\} \\
&= \frac{1}{2}(k_x^2 + k_y^2)(v_S^2 + v_A^2)\left\{1 \pm \left(\frac{k_y^2}{k_x^2 + k_y^2} + \frac{k_x^2}{k_x^2 + k_y^2}\frac{\mid v_S^2 - v_A^2 \mid^2}{(v_S^2 + v_A^2)^2}\right)^{1/2}\right\}
\end{aligned}$$

For $k_y = 0$ we recover our earlier results. For $k_y \neq 0$ we have

$$\sigma_I < \sigma_A < \sigma_{II}$$

so that the slow subspectrum, the Alfvén subspectrum and the fast subspectrum are clearly separated. The variation of $\sigma_A^2, \sigma_{sl}^2, \sigma_f^2$ as function of k_z for fixed k_x and $k_y \neq 0$ is shown on Fig. 5.5.

Let us now assume that the plasma has finite extent in the z−direction. A particularly easy case is to consider a plasma that is confined between fixed walls at $z = \pm L$ since then k_z is quantized as

$$k_z = \frac{\pi}{2L} n$$

with n an integer number that counts the number of zeroes in the z−direction. When studying MHD waves of a (1-D) nonuniform background, the number of zeroes n can be used to count the point eigenvalues.

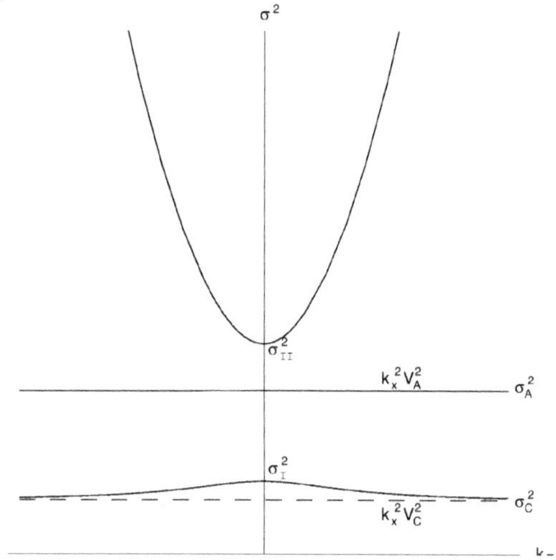

Figure 5.5: *Variation of σ^2 as function of k_z for fixed k_x and fixed $k_y \neq 0$, $v_A > v_S$.*

The basic properties of the three discrete subspectra of point eigenvalues are

- The point eigenvalue
$$\sigma_A^2 = k_x^2 v_A^2$$
of the Alfvén point spectrum is infinitely degenerated.

- The point eigenvalues of the slow subspectrum have an accumulation point for $k_z \to \infty$
$$\lim_{k_z \to \infty} \sigma_{sl}^2 = \sigma_C^2 = k_x^2 v_C^2$$

- The point eigenvalues of the fast subspectrum behave for large wavenumbers k_z as
$$\sigma_f^2 \approx k^2(v_S^2 + v_A^2), \quad \lim_{k_z \to \infty} \sigma_f^2 = \infty$$

These properties turn out the be a reliable basis for the discussion of MHD waves in a nonuniform background of finite extent. The four characteristic frequencies (5.38)

σ_C, σ_I, σ_A, σ_{II} are functions of position for an inhomogeneous background and each define an interval of frequencies. Inhomogeneity lifts the degeneracy of the point eigenvalues of the Alfvén spectrum and generates a continuum of resonant Alfvén eigenwaves. It also smears out the accumulation point of the slow spectrum into a continuum of resonant slow waves. The inequalities

$$\sigma_C < \sigma_I < \sigma_A < \sigma_{II}$$

hold at each position. Each of the four characteristic frequencies defines an interval and there four intervals can (partly) overlap. This can generate MHD waves in non-uniform plasmas with mixed character and wave transformation.

5.7 Recapitulation

"You were demonstrating in support of the other lecturers' right to a job for life?"
"Partly" said Robyn.
"But if they can't be shifted, there'll never be room for you, no matter how much better than them you may be at the job."
Vic Wilcox to Robyn Penrose,
Nice Work
D. Lodge 1988

- In this Chapter we have studied the basic MHD waves that can occur in a magnetized plasma. This has enabled us to identify v_S and v_A as the two fundamental velocities for the very-low-frequency large-scale fluid-like behaviour of plasmas. The MHD waves are studied by looking at the linear motions superimposed on a static equilibrium background. The linear motions can be studied as an eigenvalue problem, an initial value problem or a boundary value problem. We have focussed our attention on the eigenvalue problem. In ideal MHD this eigenvalue problem is Hermitian.

- In order to understand the properties of the basic MHD waves we have studied the MHD waves of a uniform plasma of infinite extent. For this background plasma we can obtain plane wave solutions and we can put the different waves in separate cages. We have used a gradual approach starting from the most simple case and putting more physics into the mathematical analysis at each step.

- The starting point is a compressible plasma in the absence of a magnetic field. The only waves present here are the sound waves. They have density and pressure variations and their displacements are longitudinal; i.e. along the wavevector $\vec{k}$. They are isotropic and driven by compression. Their phase and group velocity is the sound velocity.

- Then we remove compressibility and add a magnetic field. So, we are looking at incompressible motions in a magnetized plasma. When we restrict the motions to being

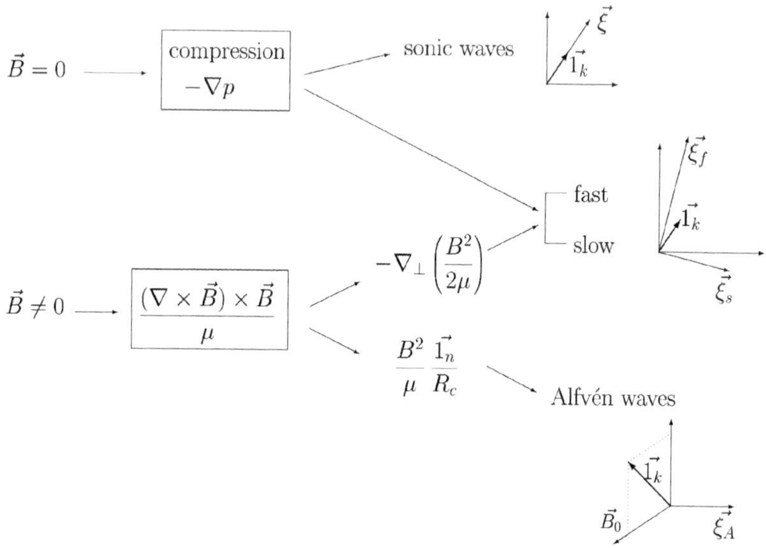

Figure 5.6: *Characterization of the different types of MHD waves in a uniform plasma*

perpendicular to the magnetic field we find the famous Alfvén waves. They are magnetic waves which leave density and pressure unchanged. They are purely driven by the magnetic tension. The displacement of Alfvén waves is polarized perpendicular to the wave vector and perpendicular to the equilibrium magnetic field. Alfvén waves are the most anisotropic animals in the zoo of MHD waves as they transport energy solely along the magnetic field lines at the Alfvén velocity.

- The plasma is still incompressible but the motions are allowed to have components perpendicular and parallel to the magnetic field lines. We now find Alfvén waves and slow waves. They have the same frequencies, but they differ essentially by their displacements. The properties of the Alfvén waves remain unchanged. Since we are incompressible on assumption, the slow waves have no density variations. They have plasma pressure variations but these are in anti-phase with and have the same amplitude as the magnetic pressure variations so that total pressure does not change. The displacement of the slow waves is confined to the plane defined by the wave vector and the equilibrium magnetic field.

- The plasma is compressible and magnetized. The Alfvén waves are still there with unchanged properties. Since we allow the plasma to be compressible we get sound waves. Since we have magnetic pressure in addition to plasma pressure we get slow and fast magnetosonic waves. In the fast magnetosonic waves plasma pressure variations and magnetic pressure variations are in phase, in the slow magnetosonic waves they are out of phase. The displacements of the slow waves and fast waves are polarized in the plane defined by the wave vector and the equilibrium magnetic field. Fast magnetosonic waves are rather isotropic; they do it a little bit better in directions perpendicular to the equilibrium magnetic field than parallel to it, but the differences are small. Slow magnetosonic waves are highly anisotropic; their transfer of energy is confined to a small cone around the magnetic field lines.

5.8 Problems

"I have a to do file?
I have a to do file???"
Homer Simpson
The Simpsons

1. Go back to the equations of ideal MHD and consider a static plasma:

$$\frac{\partial}{\partial t} = 0, \quad \vec{v} = 0$$

Convince yourself that the equation of motion is reduced to the equation of magnetostatic equilibrium:

$$-\nabla p + \frac{1}{\mu}(\nabla \times \vec{B}) \times \vec{B} = 0$$

(What has happened to the remaining equations of ideal MHD?)

2. Consider a 1-dimensional equilibrium in Cartesian coordinates (x, y, z) with

$$\vec{B} = (B_x(z), B_y(z), 0), \quad \rho(z), \quad p(z)$$

Show that the equation of magnetostatic equilibrium is

$$\frac{d}{dz}\left(p + \frac{B^2}{2\mu}\right) = 0$$

This equation tells you that total pressure has to be constant! What has happened to the magnetic tension force? Why is it absent? Note also that you have 1 equation for 4 unknown functions and that you (or better nature) are (is) left with a lot of freedom.

3. Consider a 1-dimensional equilibrium in cylindrical coordinates (r, φ, z)

$$\vec{B} = (0, B_\varphi(r), B_z(r)), \quad \rho(r), \quad p(r)$$

Show that the equation of magnetostatic equilibrium is

$$\frac{d}{dr}\left(p + \frac{B_z^2 + B_\varphi^2}{2\mu}\right) = -\frac{B_\varphi^2}{\mu r}$$

You immediately recognize the magnetic pressure force and the magnetic tension force. Again note that you have 1 equation for 4 unknown functions and that you (or better nature) are (is) left with a lot of freedom. Cylindrical 1-dimensional equilibrium models were very popular in the early days of controlled thermonuclear fusion research. Nowadays they are still used as first approximations in plasma physics and solar physics.

4. Go back to Section 5.3 on sound waves. Determine the polar diagrams for the phase velocity and the group velocity for acoustic waves! Take $v_S = 1$ in dimensionless units.

5. You are now in Section 5.4 on Alfvén waves. You are in an infinite uniform plasma with a constant magnetic field. Compute the magnetic tension force for incompressible linear plane waves with displacements perpendicular to the magnetic field lines and find that it is indeed $\vec{T}'_n = -\rho_0 \sigma_A^2 \vec{\xi}$

6. Draw the polar diagram for the phase velocity and the group velocity for Alfvén waves. Measure all velocities in the unit v_A.

7. Here is more on Alfvén waves. For linear motions superimposed on a static equilibrium the Lagrangian displacement $\vec{\xi}$ and the velocity $\vec{v}'$ are related as

$$\vec{v}' = \frac{\partial \vec{\xi}}{\partial t} = -i\sigma\vec{\xi}$$

Compute $\vec{B}'$ for incompressible linear motions in terms $\vec{\xi}$ and of $\vec{v}'$. Use this result to show that there is equipartition between the magnetic energy and the kinetic energy in the Alfvén wave or

$$\frac{\vec{B}'^2}{2\mu} = \frac{\rho_0 \vec{v}'^2}{2}$$

(R1967).

8. Use the result of the previous problem to find that the energy in the wave is

$$E_w = \rho_0 \vec{v}'^2$$

In our clever system of field aligned Cartesian coordinates (remember $\vec{B}_0 = B_0 \vec{1}_x$, $\vec{k} = k_x \vec{1}_x + k_z \vec{1}_z$) the velocity of the Alfvén wave is

$$\begin{aligned}
\vec{v}'(x, z; t) &= v_y(x, z; t)\vec{1}_y \\
&= \tilde{v}_y \cos(k_x x + k_z z - \sigma t)\vec{1}_y \\
&= \tilde{v}_y \cos\{(k_x(x - v_A t) + k_z z\}\vec{1}_y
\end{aligned}$$

Compute now the Poynting flux $\vec{P}_{\text{flux}} = \frac{1}{\mu}\vec{E} \times \vec{B} = \frac{1}{\mu} v_y B_0 \vec{1}_z \times (B_0 \vec{1}_x + \vec{B}')$, and find

$$\vec{P}_{\text{flux}} = \frac{B_0^2}{\mu} v_y \vec{1}_y + E_w v_A \vec{1}_x$$

(R1967).

9. Maybe you are not yet aware of it, but you should be puzzled by the expressions that you found for E_w and $\vec{P}_{\rm flux}$ in the previous problem. The group velocity, which is v_A for Alfvén waves, is the speed at which the wave transfers energy. This speed is just the ratio of the $\vec{P}_{\rm flux}$ to E_w. Now you see the problem. How are you going to talk you out of this situation? Look at the time dependence of the first term in the RHS of $\vec{P}_{\rm flux}$, integrate over one wave period and realize that this term is a non-persistent flux of energy (why?). Repeat the exercise for the second term and be happy. This term is the persistent flux of energy of Alfvén waves and it tells you that the energy is transported along the field lines with velocity v_A (why?). (R1967).

10. You are now in Section 5.6 on Alfvén waves and magnetosonic waves. Compute the linearized version of the Lorentz force and find

$$-\rho_0\sigma_A^2\vec{\xi}+\frac{(\vec{k}.\vec{B}_0)}{\mu}(\vec{\xi}.\vec{B}_0)\vec{k}-\frac{B_0^2}{\mu}(\vec{k}.\vec{\xi})\vec{k}+\frac{(\vec{k}.\vec{B}_0)}{\mu}(\vec{k}.\vec{\xi})\vec{B}_0$$

Compute now the magnetic pressure force and the magnetic tension force.

11. Show that the three components of the linearized equation of motion can be written as

$$\sigma^2\xi_x = k_x^2v_S^2\xi_x+k_xk_yv_S^2\xi_y+k_xk_zv_S^2\xi_z$$

$$\sigma^2\xi_y = k_xk_yv_S^2\xi_x+[k_y^2(v_S^2+v_A^2)+\sigma_A^2]\xi_y+k_yk_z(v_S^2+v_A^2)\xi_z$$

$$\sigma^2\xi_z = k_xk_zv_S^2\xi_x+k_yk_z(v_S^2+v_A^2)\xi_y+[k_z^2(v_S^2+v_A^2)+\sigma_A^2]\xi_z$$

12. Show that the three components of the linearized equation of motion can be replaced with a far simpler set of equations for the variables ξ_x, Y, Z

$$\sigma^2\xi_x-k_xv_S^2Y=0$$

$$k^2v_A^2k_x\xi_x+[\sigma^2-k^2(v_S^2+v_A^2)]Y=0$$

$$(\sigma^2-\sigma_A^2)Z=0$$

13. Go back to the eigenmode equations for Alfvén wave and magnetosonic waves. Adopt right from the start a system of field aligned Cartesian coordinates with the $x-$axis parallel to the constant magnetic field $\vec{B}_0$ and with the $z-$axis in the plane defined by the wave vector k and $\vec{B}_0$. Derive the eigenmode equations for the wave variables ξ_x, ξ_y, ξ_z. Why are these variables O.K. now?

14. Show that the frequencies σ_{sl}, σ_A and σ_f satisfy the following two sequences of inequalities

$$\sigma_C^2\le\sigma_{sl}^2\le\sigma_A^2\le k^2v_A^2\le\sigma_f^2$$

$$\sigma_C^2\le\sigma_{sl}^2\le k_x^2v_S^2\le k^2v_S^2\le\sigma_f^2$$

15. Work out the various limiting cases of subsection on eigenfrequencies and eigenvectors!

16. Show that

$$\begin{aligned}
\rho' &= -i\rho_0 Y \\
p' &= -i\rho_0 v_S^2 Y \\
\vec{B}' &= ik_x B_0 \vec{\xi} - i\vec{B}_0 Y \\
p'_m &= \frac{\vec{B}'.\vec{B}_0}{\mu} = ik_x \rho_0 v_A^2 \xi_x - i\rho_0 v_A^2 Y \\
p'_t &= ik_x \rho_0 v_A^2 \xi_x - i\rho_0 (v_S^2 + v_A^2) Y \\
\frac{1}{\mu}(\vec{B}_0.\nabla)\vec{B}' &= ik_x p'_m \vec{1}_x - \rho_0 \sigma_A^2 (\xi_y \vec{1}_y + \xi_z \vec{1}_z) \\
\vec{T}'_n &= -ik_x p'_m \vec{1}_x + \frac{1}{\mu}(\vec{B}_0.\nabla)\vec{B}' = -\rho_0 \sigma_A^2 (\xi_y \vec{1}_y + \xi_z \vec{1}_z)
\end{aligned}$$

17. Show that you can rewrite the eigenvalue equation (5.27) for magnetosonic waves as

$$\frac{k_z^2 v_S^2}{\sigma_{sl,f}^2 - k_x^2 v_S^2} = \frac{\sigma_{sl,f}^2 - k^2 v_A^2}{\sigma_{sl,f}^2}$$

18. Determine $\vec{\xi}_{sl}$ and $\vec{\xi}_f$ and show that $\vec{\xi}_A$, $\vec{\xi}_{sl}$ and $\vec{\xi}_f$ are three orthogonal vectors.

19. Show that for magnetosonic waves

$$\begin{aligned}
\vec{B}' &= iB_0 \xi_z (-k_z \vec{1}_x + k_x \vec{1}_z) \\
p' &= -i\rho_0 v_S^2 Y \\
p'_m &= i\rho_0 v_A^2 \frac{k_x^2 v_S^2 - \sigma_{sl,f}^2}{\sigma_{sl,f}^2} Y \\
\frac{p'_m}{p'} &= \frac{v_A^2}{v_S^2} \frac{\sigma_{sl,f}^2 - k_x^2 v_S^2}{\sigma_{sl,f}^2} = \frac{k_z^2 v_A^2}{\sigma_{sl,f}^2 - k^2 v_A^2} \\
\frac{p'_t}{p'} &= \frac{v_S^2 + v_A^2}{v_S^2} \frac{\sigma_{sl,f}^2 - k_x^2 v_C^2}{\sigma_{sl,f}^2} = \frac{\sigma_{sl,f}^2 - k_x^2 v_A^2}{\sigma_{sl,f}^2 - k^2 v_A^2} \\
\vec{T}'_n &= -\rho_0 \sigma_A^2 \xi_z \vec{1}_z
\end{aligned}$$

20. Determine the polar diagrams of the phase velocity for the slow and fast magnetosonic waves!

21. Compute and subsequently draw the variation of $\sigma_A^2, \sigma_{sl}^2, \sigma_f^2$ as function of k_x for fixed k_z. In particular determine the values of σ_{sl}^2, σ_f^2 for $k_x = 0$ and for $k_x \to \infty$. Now repeat this exercise with k_x replaced with k_z and vice versa. Compute and subsequently draw the variation of $\sigma_A^2, \sigma_{sl}^2, \sigma_f^2$ as function of k_z for fixed k_x. In particular determine the values of σ_{sl}^2, σ_f^2 for $k_z = 0$ and for $k_z \to \infty$.

22. Obtain the following expressions for the components of the group velocity of the slow and fast magnetosonic waves

$$v_{gr,x} = v_{ph} \cos\theta \mp \frac{v_C^2 \cos\theta \sin^2\theta}{v_{ph}\,(1-\Delta)^{1/2}}, \quad v_{gr,z} = v_{ph} \sin\theta \pm \frac{v_C^2 \cos^2\theta \sin\theta}{v_{ph}\,(1-\Delta)^{1/2}}$$

where

$$\Delta(\theta) = 4\frac{v_C^2}{v_A^2 + v_S^2}\cos^2\theta$$

Verify that these expressions are indeterminate for the slow wave when $\theta \to \pi/2$ since $\lim_{\theta\to\pi/2} v_{ph} = 0$.

Now lift this indeterminacy by rewriting the expressions as

$$\begin{aligned} v_{gr,x} &= v_{ph}\cos\theta \mp \frac{v_C}{\sqrt{2}} \frac{[1 \mp (1-\Delta)^{1/2}]^{1/2}}{(1-\Delta)^{1/2}} \sin^2\theta \\ v_{gr,z} &= v_{ph}\sin\theta \pm \frac{v_C}{\sqrt{2}} \frac{[1 \mp (1-\Delta)^{1/2}]^{1/2}}{(1-\Delta)^{1/2}} \sin\theta\cos\theta \end{aligned}$$

23. Show that

$$v_{gr,z}k_z > 0 \text{ for fast waves,} \quad v_{gr,z}k_z < 0 \text{ for slow waves}$$

24. Determine the polar diagrams of the group velocity for the slow and fast magnetosonic waves!

25. Compute the displacement for the slow wave in the limit $k_z \to \infty$ and find that the slow wave in this limit is polarized parallel to the magnetic field lines.

26. Consider a coronal loop with $B_0 = 10$ Gauss $= 10^{-3}$ Tesla, $L = 5 \times 10^7$ m and $n = 5 \times 10^{14}$ m^{-3}, $\rho_0 = 8 \times 10^{-8}$ kg m^{-3}. Compute for these values the Alfvén speed and find that it is approximately $v_A = 10^6$ m s^{-1}. Compute the period of the fundamental Alfvén oscillation mode of this loop. Use $k = 2\pi/L$. (H2000).

Hannes Alfvén (1908 - 1995)
During his career, Alfvén made a number of fundamental theoretical discoveries. The one for which he is best known is the magnetohydrodynamic (hydromagnetic) wave, commonly called the Alfvén wave. But he invented a number of other fundamental concepts that are not so closely associated with his name. These include simplifications in the concepts with which we treat the behavior of ionized gases (plasmas). He found the established way of calculating particle orbits (Strmer orbit theory) to be impractical, especially in the energy range relevant to auroras. This led him to develop, as a tool, the guiding-center approximation for the motion of charged particles in electric and magnetic fields. He also discovered the first adiabatic invariant of charged particle motion, and he invented the concept of frozen-in magnetic flux. Together, these tools established magnetohydrodynamics as a resource and as a field of research. It is hard to imagine working today in plasma physics without using the tools he provided us. This work, cited as "contributions and fundamental discoveries in magnetohydrodynamics" earned him a Nobel Prize for Physics.
Hannes Alfvén possessed a gift that allowed him to extract results of great importance and generality from specific problems. It is a mark of his genius that his initial understanding came primarily from physical reasoning; the mathematical demonstrations came only after he had, in his mind's eye, determined the physical process. The discovery of Alfvén waves is, in many ways, representative of his approach. It grew out of a specific problem, namely that of sunspots. He first determined that it was possible to propagate electromagnetic waves in a highly conducting plasma. (This he claimed was the easy part.) Only then did he develop the mathematical demonstrations. The idea that such waves were possible ran contrary to the conventional thought of the time because it was taught that electromagnetic waves could propagate no more than a skin depth (about a wavelength) into a good conductor. But Alfvén had found, by pure power of intellect, an entirely new propagation mode. He discovered how electromagnetic waves can propagate without damping in a plasma of arbitrarily high conductivity.
Hannes Alfvén received the Nobel Prize in Physics in 1970 from the Swedish King Gustavus Adolphus VI.
It usually took years for his ideas to be accepted. For example, his discovery of hydromagnetic waves was presented in an admirably simple and clear mathematical form in a Letter to Nature published in 1942. Acceptance came suddenly some six years later when, as Alfvén recounted, at the end of a seminar he gave at the University of Chicago in 1948, the famous physicist Enrico Fermi nodded his head and said "of course" and, according to Alfvén's account, the prestige of Fermi was such that "the next day, everyone nodded and said, 'of course'".

C.-G. Fälthammar, Royal Institute of Technology, Stockholm. A. J. Dessler, University of Arizona, Tucson

Chapter 6

The solar wind

> And this I believe: that the free, exploring mind of the individual human is the most valuable thing in the world. And this I would fight for: the freedom of mind to take any direction it wishes, undirected. And this I must fight against: any idea, religion, or government which limits or destroys the individual.
> *East of Eden*
> J. Steinbeck

The prediction by E. Parker in 1958 (Parker, E.N.: 1958, Ap.J. **128**, 664) of a high speed solar wind with supersonic velocities of several hundreds of km s^{-1} at the Earth's orbit caused a huge controversy in the astrophysical and geophysical communities. Parker's theory was received with a great deal of skepticism, even disbelief. Fortunately for E. Parker, Russian and American satellites soon afterwards did indeed discover this fast streaming solar wind. By the end of the 1960's the solar wind was an established astrophysical fact and since then it has become an integral and natural part of solar physics and astrophysics.

This Chapter follows the historic route of the advancement of our understanding of the solar wind. The reader will realize that the process of scientific discovery is often very similar to the individual learning process of the average undergraduate student. At first, we do not know anything about the subject or we are on the wrong track. Then we understand a little piece of the puzzle and are very happy and often also very pleased with ourselves. However, soon we do realize that it is indeed just a little piece and we try to take the next step. Usually we fail and need several iterations before we get it right.

6.1 Overview of observations

It is only in the late 1930's that the solar corona was unambiguously recognized as being composed of very hot plasma. The key in reaching this conclusion was the realization that many of the unidentified lines seen in coronal spectra were not due to unknown elements, but belonged to high ionization stages of known elements. Our understanding of the structure of

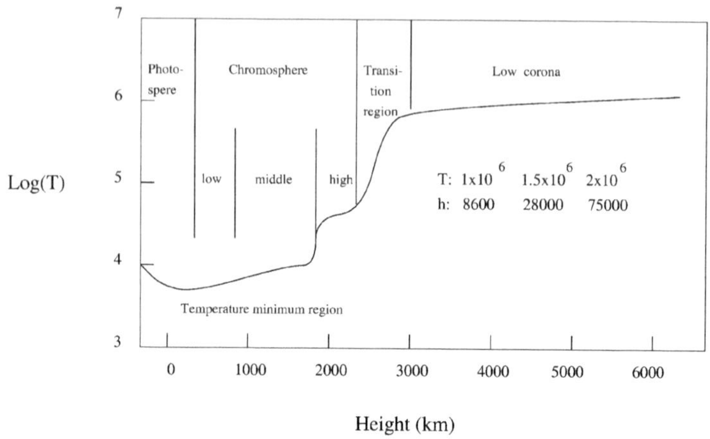

Figure 6.1: *The temperature profile in the solar atmosphere.*

the slowly varying part of the corona comes from spectroscopy. By modelling radiative transfer in atomic spectral lines at different depths in the solar atmosphere and the lower corona, it is possible to construct a fairly accurate representation of density, pressure and temperature as functions of height above the solar photosphere. The variation of temperature with height, as shown in Fig. 6.1, is spectacular with the transition from the photospheric temperature of about $6000K$ to the high coronal temperatures of several $10^6 K$ occurring over a distance of a few hundreds of km (this region is aptly called the transition region). Because of its high temperature, the corona suffers tremendous energy losses. These include losses due to thermal conduction, radiation and kinetic energy carried away by mass outflows. For a more or less steady state to be maintained, an energy input at the base of the corona of the order of a few $10^{21} W$ is required. This does not seem to be a big deal taking into account that the solar luminosity is $3.86 \times 10^{26} W$. We only need to divert, in a more or less continuous fashion, something like 0.001% of the total. However, the fundamental problem is that we are required to draw energy from a reservoir (the solar surface) to heat up the corona at a higher temperature than the reservoir itself!

The corona is far from a static structure. The concept of an average corona at a given time is helpful in describing the large scale global behaviour of the corona. Close observation reveals temporal and spatial variations on almost every scale investigated. Dramatic changes in the large-scale structure of the corona occur in parallel with the 11 yr solar cycle (see Fig. 1.5). At solar maximum, the corona shows an overall spherical symmetry or rather a spherical distribution of helmet streamers. At solar minimum, the solar corona is no longer spherically symmetric, although axial symmetry with respect to the rotation axis still exists. The most prominent feature of the corona at solar minimum is the presence of coronal holes over the higher latitudes in both the north and south solar hemispheres. Helmet streamers are now restricted to a latitudinal band extending to $30° - 40°$ above the equator. As the sun proceeds from solar minimum to solar maximum, the equatorial band expands in latitude, gradually filling the coronal holes. The coronal holes are not empty; they appear darker because the particle density is lower there than in the brighter active equatorial regions.

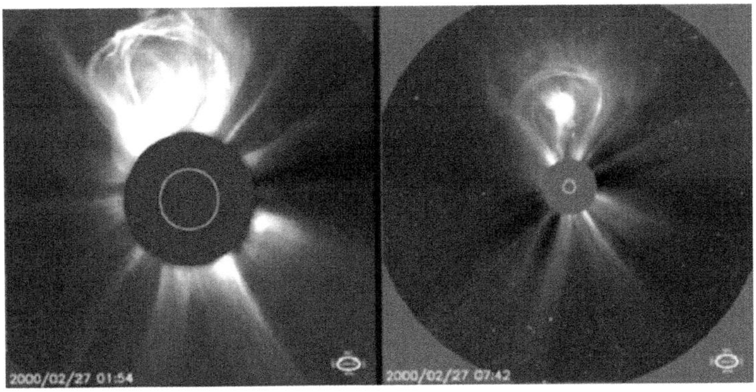

Figure 6.2: *"Lightbulb" coronal mass ejection on February 27, 2000 observed by LASCO. (Courtesy of SOHO/LASCO consortium. SOHO is a project of international cooperation between ESA and NASA.)*

At solar minimum, the large scale magnetic field close to the sun is roughly dipolar, and maps reasonably well on the corona. Coronal holes correspond to regions of open magnetic field lines, i.e. field lines that extend far out into the interplanetary space. The equatorial, active regions coincide with the closed field line regions, i.e. the regions of the corona pervaded by magnetic field lines that originate in one hemisphere, cross the equatorial plane to anchor themselves in the other hemisphere, at latitudes comparable to their point of origin.

The corona is also the host of a variety of phenomena appearing and evolving on shorter time scales: flares, eruptions, coronal mass ejections, etc. The most spectacular short-lived events are coronal mass ejections. An example is shown on Fig. 6.2. These are explosive like events in which, over timescales of hours, a large cavity, of linear dimension of $1R_\odot$, is blown out of the corona. Matter is ejected at speeds in the range $10-2000$ km/s. The mass involved is in the range of $10^{11} - 10^{13}$ kg. The total energy required to lift this mass out of the sun's gravitational field to the observed velocities is $10^{24} - 10^{25}$ J. Unbelievable! An example of a coronal mass ejection (CME) is shown on Fig. 6.2. It is borrowed from the "Best of SOHO Images - Latest Picks" at http://sohowww.nascom.nasa.gov/

Although it had been realized earlier in this century that episodic mass outflows from the sun are responsible for aurorae and geomagnetic storms, evidence for the presence of a continuous corpuscular outflow from the sun was only put forth in the 1940's and 1950's. The first and clearest pre-satellite evidence for a continuous particle stream emanating from the sun came from the ionic tails from comets. These tails always point away from the sun. In 1943 Hoffmeister found that there is a small systematic deviation: the tails axis slightly lags (in the sense of the comet's motion) behind the prolonged radius vector, so that there is a small angle ϵ, normally $< 5°$, between the tail axis and the solar radius vector. The explanation is entirely analogous to Bradley's explanation of stellar aberration. The orientation of the ion tail is determined by the relative vector velocities of the radially expanding solar wind plasma and the comet in its orbit. The angle ϵ, the velocity of the comet perpendicular to the radius vector $v_\perp$ and the radial velocity of the solar wind v_{wind}, is

$$\tan\epsilon = \frac{v_\perp}{v_{\text{wind}}} \tag{6.1}$$

Analysis of Hoffmeister's data gives an average value of the solar wind speed of 474 ± 21 km/s. It is curious that Hoffmeister did not calculate the solar wind velocity from his data. Biermann in the 1950's realized that the solar radiation pressure is far too small to push the comet tails away from the sun. He suggested instead that there must be particles streaming from the sun, which would push the evaporated gases outward and produce the oriented tails. With this postulated corpuscular radiation Biermann was able to explain Hoffmeister's discovery of the relation between the angle ϵ, $v_\perp$ and v_{wind}. From the distribution of comets it could be deduced that the hypothetical corpuscular radiation blows at all times (continuously) and in essentially all directions. What is more, the large variations in space and time of the phenomenon could already be seen by its variable effects on comets.

In 1957 Chapman considered a "new" model of the solar corona. Chapman assumed that the corona and its distant extensions were static and that energy was transferred only by conduction. Chapman's model made two important points: (i) conduction is important for maintaining the corona and its extension; (2) the gaseous medium near the earth is regarded as an extended part of the corona. It was not clear how Biermann's hypothetical corpuscular radiation and Chapman's extended atmosphere were related. Parker found that in Chapman's static model the density actually goes through a minimum and then increases and that pressure remains finite as $r \to \infty$ and is much larger than any reasonable interstellar pressure which might be invoked for balance. Parker allowed for a stationary rather than a static equilibrium and concluded that the particles not only could stream from the sun but that they necessarily do so. The corona must expand! The particles are nothing less than the corona itself. Biermann's hypothetical corpuscular radiation and Chapman's extended atmosphere are the same object, only that the extended atmosphere is not static. Parker's hydrodynamic theory contains *a transonic solution* which reproduces the velocities required by Biermann's comet work and has zero pressure at large distances. Parker christened this phenomenon the *solar wind.* Parker's solution predicts a solar wind speed at the earth's orbit of the order of a few hundreds of km/s, which is definitely supersonic. This implies that the solar wind undergoes a sonic transition between the solar surface and the earth's orbit. In 1962 Chamberlain argued that the solar expansion is not a wind but a *breeze* with velocities that are subsonic everywhere! Parker also calculated that the solar wind and solar rotation would draw out the magnetic field lines into an *Archimedean spiral* (see Fig. 1.6).

Observations from space are necessary to measure the wind speed directly, because the magnetic field of the earth act as a buffer against charged particles, and essentially no information about the flow of coronal particles trickles down to the surface of the earth. The first in situ measurements of the solar wind wind were made by a group of Russian scientists with instruments on board Lunik III and Venus I. The first American solar wind measurements were carried out by an M.I.T. group on board Explorer 10 in 1961. All reasonable doubt concerning the existence of an essentially continuous solar wind was removed in 1962 by measurements made on board the Venus probe, Mariner 2. Approximately three continuous months of data were obtained.

The principal characteristics of the solar wind are:

- a detectable high speed (supersonic), continuous, but rather variable, outflow is present at all times. It is definitely a wind not a breeze!

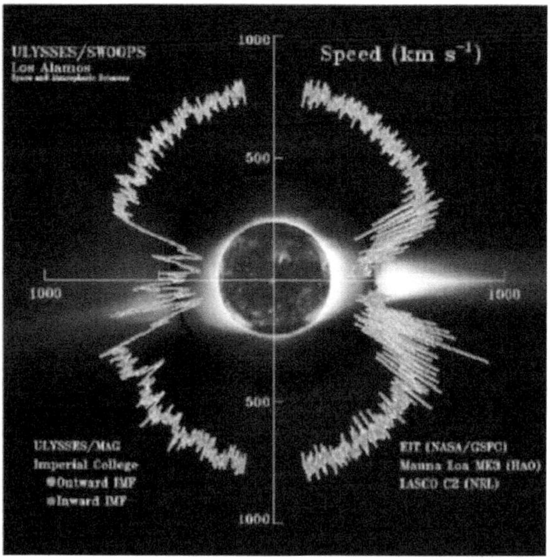

Figure 6.3: *The variable speed of the solar wind.*

- electrons, protons and $\alpha-$particles $(3-4\%$ relative to the protons) are the predominant constituents
- the average speed is 500 km/s but the speed varies between 300 km/s and 800 km/s. The solar wind consists of two separate components: the low-speed wind of high density and the high-speed wind of low density as is clearly shown on Fig. 6.3. Fig. 6.3 is an overlay plot of SWOOPS solar wind speed data and EIT/LASCO/Mauna Loa images of the solar corona. It can be found at http://helio.esa.int/ulysses and at http://swoops.lanl.gov/lasco_Swoops.html (SOHO is a project of international cooperation between ESA and NASA.)
- the high speed solar wind streams originate from polar coronal holes and/or at times of solar maximum, from smaller holes in the equatorial active regions, where the magnetic field lines are open. Low speed streams emanate from closed field line regions. The low-speed wind is remarkably constant in times of reduced magnetic activity .
- the high speed streams have a strong tendency of recurring every 27 days, corresponding to the solar rotation period.

The interplanetary magnetic field has been measured and the configuration found to fit (on average) Parker's description of an Archimedean spiral by the Interplanetary Magnetic Probe (IMP) in 1963. The magnetic field has been found to be closely related to the photospheric field. The magnetic field couples the solar wind plasma to the sun and results in a loss of solar angular momentum sufficient to be important in the structure and evolution of the sun.

Transient events associated with the passage of MHD shocks at the earth's orbit are superimposed on both the slow-speed and the high-speed wind. These transient events can be traced back to eruptions, flares and coronal mass ejections in the lower corona. These energetic disturbances are advected outward by the wind and steepen into shocks. Their interaction with the earth's magnetosphere causes the aurorae and the geomagnetic storms.

6.2 Alternatives to the energy equation

The set of MHD equations is extremely difficult even for relatively simple geometries and magnetic field configurations. A popular avenue of simplification in theoretical solar wind research has been to concentrate on the dynamics of the flow and to make rather drastic assumptions regarding the energetics of the flow. In plain English, to get rid of the energy equation. But if the energy equation is omitted from the set of MHD equations, we need an additional relation between some of our system variables in order to still have a closed set of equations (or as many equations as unknown variables). In solar wind theory *Bernoulli's equation* turns out to a key equation. Let us go back to elementary fluid mechanics and recall Bernoulli's equation for a stationary flow in a vertical gravity field

$$\frac{v^2}{2} + \int_{\rho_0}^{\rho} \frac{dp}{\rho} + gz = \text{ constant along a flow line} \tag{6.2}$$

The terms in Bernoulli's equation are kinetic energy per unit mass, enthalpy per unit mass and potential energy per unit mass. The second term complicates the analysis because of the integral. In realistic situations this integral cannot be written in simple closed analytical form. In order to make analytical progress the obvious assumption is that $p = f(\rho)$. The two classic assumptions in solar wind theory are isothermal and polytropic plasmas. The plasma or flow is said to be isothermal if

$$T = \text{constant} \tag{6.3}$$

and polytropic if

$$p = K\rho^{\alpha}, \quad K = \text{constant}, \quad \alpha = \text{constant} \tag{6.4}$$

The classic speed of sound is defined by (4.14). In this definition γ is the ratio of the specific heats at constant pressure and constant volume respectively $\gamma = c_P/c_V$. With the introduction of isothermal and polytropic plasmas it is convenient to define two additional versions of the speed of sound : i.e. the isothermal speed of sound, $v_{S,i}$ and the polytropic speed of sound $v_{S,p}$. They are defined as

$$v_{S,i}^2 = \frac{p}{\rho} = \frac{\overline{\mathcal{R}}}{\overline{\mu}}T = \frac{v_S^2}{\gamma}, \quad v_{S,p}^2 = \frac{\alpha p}{\rho} = \frac{\alpha}{\gamma}v_S^2 \tag{6.5}$$

For an isothermal plasma T is constant. In that case $v_{S,i}^2$ is also constant and

$$\frac{dp}{d\rho} = v_{S,i}^2 \, , \quad \int \frac{dp}{\rho} = v_{S,i}^2 \ln \rho \tag{6.6}$$

For an polytropic plasma

$$\frac{dp}{d\rho} = v_{S,p}^2 \, , \quad \int \frac{dp}{\rho} = K\frac{\alpha}{\alpha - 1}\rho^{\alpha - 1} \tag{6.7}$$

Isothermal flow is a special case of polytropic flow. It occurs when $\alpha = 1$. Often the index α is chosen equal to γ.

6.3 Static models

Let us go back in time to the 1940's and the 1950's and be static! The astrophysical community is convinced that the corona is indeed extremely hot and temperatures of the order of million K are accepted. The mechanism for coronal heating is not known, but we assume that the heat is deposited at some level r_0 in a spherically symmetric corona. This heat input is modelled by prescribing the temperature T_0 at the level $r = r_0$. The aim is then to construct a model for the corona above (and also below) this reference level, at which the density (ρ_0), pressure (p_0) and temperature (T_0) are known. OK, let us start with a simple model. Simple means that we want the model to be as follows.

- It is in static equilibrium. The model is time independent and there is no velocity field. In particular we forget that the sun rotates.
- It is spherically symmetric. In a system of spherical polar coordinates (r, θ, φ) all quantities only depend on the radial distance to the centre of the sun. The only derivative that survives is d/dr.
- There is no magnetic field.

The equation of hydrostatic equilibrium is

$$-\frac{1}{\rho}\frac{dp}{dr} - \frac{G\mathcal{M}_\odot}{r^2} = 0 \tag{6.8}$$

We use the gas law (3.40) to write density in terms of pressure and temperature $\rho = (\overline{\mu}/\overline{\mathcal{R}})(p/T)$. We then insert this expression for ρ in the equation of hydrostatic equilibrium and obtain by integration the following result for pressure

$$p(r) = p_0 \exp\left\{-G\mathcal{M}_\odot \frac{\overline{\mu}}{\overline{\mathcal{R}}} \int_{r_0}^{r} \frac{dr}{r^2 T}\right\} \tag{6.9}$$

For the present discussion we prefer analytical expressions for the variation of pressure and density with distance r. Hence we restrict our efforts to profiles for $T(r)$ that enable us to rewrite the integral involving $T(r)$ in terms of simple functions. Let us get very simple and become isothermal.

Isothermal static corona

Since $T(r) = T_0$ and

$$\int_{r_0}^{r} \frac{dr}{r^2 T} = \frac{1}{T_0 r_0}\left(1 - \frac{r_0}{r}\right)$$

it follows from (6.9) and (3.40) that pressure and density vary with r as

$$\begin{aligned} p(r) &= p_0 \exp\left\{-\frac{G\mathcal{M}_\odot}{r_0 v_{s,i}^2}(1 - \frac{r_0}{r})\right\} \\ \rho(r) &= \rho_0 \exp\left\{-\frac{G\mathcal{M}_\odot}{r_0 v_{S,i}^2}(1 - \frac{r_0}{r})\right\} \end{aligned} \tag{6.10}$$

The static coronal model has to blend into the interstellar medium implying that pressure and density must match the interstellar density and interstellar pressure for very large heliospheric distances or formally for $r \to \infty$. Now

$$p_\infty = p_0 \exp\left\{-\frac{G\mathcal{M}_\odot}{r_0 v_{S,i}^2}\right\}, \quad \rho_\infty = \rho_0 \exp\left\{-\frac{G\mathcal{M}_\odot}{r_0 v_{S,i}^2}\right\}$$

Neither pressure nor density vanish at infinity. Since the isothermal model is a very crude approximation of reality we agree to postpone comparison with the interstellar medium values to the following subsection.

The static heat-conduction corona

With the acceptance of the high coronal temperatures of the order of million K in the 1940's came also the recognition that the thermal conductivity of the coronal plasma, which is proportional to $T^{5/2}$, implies very efficient outward heat transport from the inner corona. In 1957 Chapman considered a "new" model of the corona. It was assumed that the corona and its distant extensions were static and that energy was transferred only by thermal conduction; all other sources of energy transport such as radiative losses were assumed to negligible. We do not worry about how the corona is heated, we just take it for granted that it is there. For simplicity we assume that the heat is deposited at a level r_0 in a spherically symmetric corona and we model this heat input by prescribing the temperature at this level $T(r_0) = T_0 = 1 - 2 \times 10^6$K. The temperature distribution is determined by the conservation of conductive flux (energy equation)

$$\nabla.(\kappa \nabla T) = \frac{1}{r^2}\frac{d}{dr}\left(r^2 \kappa \frac{dT}{dr}\right) = 0 \tag{6.11}$$

This can be integrated to

$$r^2 \kappa \frac{dT}{dr} = \text{constant}$$

In a high temperature plasma of fully ionized hydrogen, an approximate but fairly accurate expression for the thermal conductivity is $\kappa(T) = \kappa_0 T^{5/2}$. With the use of this expression for κ we can rewrite the equation for T as

$$r^2 T^{5/2} \frac{dT}{dr} = \text{constant}$$

This equation governs the radial variation of temperature T due to conduction from $R_\odot$ up to $+\infty$. Since $T(\infty) << T_0$ it is a good approximation to put $T(\infty) = 0$. Similarly, since $T_\odot = T(R_\odot) \approx 6 \times 10^3$ K $<< T_0 = 1 - 2 \times 10^6$ K, the error introduced for the transition region and the corona by setting $T_\odot = 0$ is negligible. The solution is

$$\begin{aligned} T(r) &= \left\{T_\odot^{7/2}\frac{r_0/r - 1}{r_0/R_\odot - 1} + T_0^{7/2}\frac{R_\odot/r - 1}{r_0/R_\odot - 1}\right\}^{2/7} \quad \text{for } r \le r_0 \\ T(r) &= T_0 \left(\frac{r_0}{r}\right)^{2/7} \quad \text{for } r \ge r_0 \end{aligned} \tag{6.12}$$

The heat-conduction temperature profile (6.12) models rather accurately the extremely steep gradient in the transition region and the slow decline towards interplanetary space. For $r_0 = 1.2R_\odot$ and $T_0 = 1 - 2 \times 10^6$K, the temperature at the earth is in excess of 10^5K so that the earth is enveloped in a hot plasma. Let us have a look at pressure. Since

$$\int_{r_0}^{r} \frac{dr}{r^2 T} = \frac{7/5}{T_0 r_0}\left(1 - (\frac{r_0}{r})^{5/7}\right)$$

it follows from (6.9) and (3.40) that pressure and density vary with r as

$$\begin{aligned} p(r) &= p_0 \exp\left\{-\frac{7/5 G\mathcal{M}_\odot}{r_0}\frac{\overline{\mu}}{\mathcal{R}T_0}(1 - (\frac{r_0}{r})^{5/7})\right\} \\ \rho(r) &= \rho_0 \left(\frac{r}{r_0}\right)^{2/7} \exp\left\{-\frac{7/5 G\mathcal{M}_\odot}{r_0}\frac{\overline{\mu}}{\mathcal{R}T_0}(1 - (\frac{r_0}{r})^{5/7})\right\} \end{aligned} \qquad (6.13)$$

Pressure decreases monotonically with increasing r approaching the finite value

$$p(\infty) = p_0 \exp\left\{-\frac{7/5 G\mathcal{M}_\odot}{r_0}\frac{\overline{\mu}}{\mathcal{R}T_0}\right\}$$

The question is whether this extended static Chapman coronal model can blend into the interstellar background for large heliospheric distances or for $r \to \infty$. For $r_0 = 1.2R_\odot$, $T_0 = 1 - 2 \times 10^6$K, $p_0 = 10^{-3}$Pa it follows that $p_\infty = 10^{-8}$Pa. This value should be compared with the pressure in the interstellar medium $p_{ISM} = 10^{-15}$Pa. This is bad. The pressure that the interstellar medium can provide is insufficient by 7 orders of magnitude. If you are not yet impressed by this huge mismatch for pressure, it gets worse! Look at the density $\lim_{r\to\infty} \rho(r) = +\infty$! The conclusion is obvious. The heat conduction corona cannot be embedded in interstellar space under pressure equilibrium. Although Chapman's model is obviously wrong, it made two fundamental points:

- it emphasizes the importance of thermal conduction in the maintenance of the corona and its extension
- it demonstrates that the corona cannot terminate near the sun; rather coronal (and thus solar) material must extend far out into interplanetary space. In particular, the gaseous medium near the earth is simply an extended part of the solar corona.

Think harder!

We are now convinced that the static coronal model is in deep trouble. What has gone wrong? It is high time to move on and become stationary. This becomes very clear by looking at Bernoulli's equation (this is one our favorite equations in the following sections, so make sure you like it or you will have a though time):

$$\frac{v^2}{2} + \int_{p_0}^{p} \frac{dp}{\rho} - \frac{G\mathcal{M}_\odot}{r} = E \qquad (6.14)$$

with E a constant. Let us keep it simple and consider a polytropic flow (and take $\alpha = \gamma$) so that

$$\begin{aligned}\frac{v^2}{2} + \frac{\gamma p}{(\gamma-1)\rho} - \frac{G\mathcal{M}_\odot}{r} &= E \\ \frac{v^2}{2} + \frac{\gamma}{(\gamma-1)}\frac{\overline{\mathcal{R}}}{\overline{\mu}}T - \frac{G\mathcal{M}_\odot}{r} &= E\end{aligned}$$

What are we trying to do here? Well, we would like to relate conditions at large heliospheric distances (formally $r \to \infty$) to conditions in the lower corona at $r = r_0$. For that reason we rewrite Bernoulli's equation as:

$$\underbrace{\frac{v_0^2}{2}}_{\approx 0} + \frac{\gamma}{(\gamma-1)}\frac{\overline{\mathcal{R}}}{\overline{\mu}}T_0 - \frac{G\mathcal{M}_\odot}{r_0} = \frac{v_\infty^2}{2} + \frac{\gamma}{(\gamma-1)}\frac{\overline{\mathcal{R}}}{\overline{\mu}}\underbrace{T_\infty}_{\approx 0} - \underbrace{\lim_{r\to\infty}\frac{G\mathcal{M}_\odot}{r}}_{=0}$$

or

$$\frac{\gamma}{(\gamma-1)}\frac{\overline{\mathcal{R}}}{\overline{\mu}}T_0 - \frac{G\mathcal{M}_\odot}{r_0} \approx \frac{v_\infty^2}{2}$$

In its extreme simplicity, this is a key equation. It tells us that there is a transfer of enthalpy to kinetic energy and that the corona must expand if it is sufficiently hot, which is the case if

$$\frac{\gamma}{(\gamma-1)}\frac{\overline{\mathcal{R}}}{\overline{\mu}}T_0 > \frac{G\mathcal{M}_\odot}{r_0} \qquad (6.15)$$

Here are some comments for the literally inclined reader! At a temperature as high as $1 - 2 \times 10^6$K, the pressure and density ought to remain high for a distance of several solar radii above the solar surface. The solar gravity, which binds the corona to the sun, weakens as the square of the distance, so the ability of the sun to hold on to its corona decreases at great distances. In fact, at very large distances, gravity by itself is too weak to prevent the corona from rushing outward, and it could be kept in place only by the additional pressure of the interstellar gas, pressing inward. Yet because of the high temperature of the corona, its pressure overwhelms the restraining pressure of the interstellar gas. The only conclusion is that the corona cannot be static but must expand continuously outward into the near vacuum between the stars.

6.4 de Laval nozzle

Why do we bother with this Section? The reason is that the process by which the solar corona with typical thermal speeds of 100 - 250 km/s is able to move material against the solar gravitational field (such that the escape velocity is $\approx$ 600km/s) and is able to give it a velocity of 400 - 500 km/s may at first sight seem somewhat mysterious. Values of thermal speeds can be calculated using the expressions for the isothermal sound velocity (6.5) $v_{S,i} = (\overline{\mathcal{R}}T/\overline{\mu})^{1/2}$, the classic sound velocity (4.14) $v_S = (\gamma p/\rho)^{1/2}$, the thermal proton velocity (2.16) $v_{t,p} = (\overline{\mathcal{R}}T)^{1/2}$, or its variant $\tilde{v}_{t,p} = (3\overline{\mathcal{R}}T)^{1/2}$. For $\overline{\mu} = 1/2$ and $T_0 = 1 - 2 \times 10^6$ K we get $v_{S,i} = 130-180$ km/s, $v_S = 170-230$ km/s, $v_{t,p} = 90-130$ km/s, $\tilde{v}_{t,p} = 155-225$ km/s. The escape velocity of the sun is $v_{esc,\odot} = (2G\mathcal{M}_\odot/R_\odot)^{1/2}$. With $G = 6.6726 \times 10^{-11}$ m^{3}s^{-2}kg^{-1},

$\mathcal{M}_\odot = 1.99 \times 10^{30}$ kg and $R_\odot = 6.96 \times 10^8$ m we get $v_{esc,\odot} = 620$ km/s. Doppler shifts of coronal radar echoes indicate outward motions at a height of $1.3R_\odot$ of $10 - 20$ km/s. The previous subsection told us that specific enthalpy is converted into kinetic energy of expansion but we want to have a better understanding of the basic physics of this accelerating process. Here we can learn from aircraft engineers who know about an analogous accelerating process for a rather long time.

Let us look at the flow of a fluid down an axisymmetric tube, oriented parallel along the $x-$axis, with a cross-section σ which is a function of x (the classic name for this system is nozzle). Here you might think of a hose-pipe for watering your lawn and you know that by reducing σ you increase the speed of the outflowing water and conversely that by increasing σ you decrease the speed of water. Is that the only kind of flow behaviour you can think of? Think of traffic on a highway. In every cardriver's mind the word bottleneck is associated with slow traffic and low speeds! When a lane is closed for construction on a highway the traffic slows down and eventually comes to a standstill. When the lane is opened again you can see the cars speeding away at high velocity. So what is the difference between these two kinds of flow behaviour?

Incompressible flow: $\rho =$ constant

Let us look at an incompressible fluid that is forced to flow in the nozzle. In a steady state ($\frac{\partial}{\partial t} = 0$) the mass conservation equation tells us that

$$\sigma(x)v_x(x) = C_1$$

where C_1 is a constant. This simple constraint implies that

$$\text{converging tube}: \frac{d\sigma}{dx} < 0 \quad \rightarrow \quad \frac{dv_x}{dx} > 0 : \text{flow accelerates}$$

$$\text{tube with constant cross-section}: \frac{d\sigma}{dx} = 0 \quad \rightarrow \quad \frac{dv_x}{dx} = 0 : \text{constant flow}$$

$$\text{diverging tube}: \frac{d\sigma}{dx} > 0 \quad \rightarrow \quad \frac{dv_x}{dx} < 0 : \text{flow decelerates}$$

The flow accelerates in a converging tube and decelerates in a diverging tube.

Compressible adiabatic flow

Now we force a compressible fluid to flow in the same nozzle.

• **Mass conservation**

The equation of mass conservation tells us that

$$\rho(x)\sigma(x)v_x(x) = C_1$$

where C_1 is a constant. Differentiation yields

$$\frac{1}{\rho}\frac{d\rho}{dx} = -\frac{1}{\sigma}\frac{d\sigma}{dx} - \frac{1}{v_x}\frac{dv_x}{dx} \tag{6.16}$$

• **Equation of motion**

For a stationary flow along the x direction and in absence of gravity the equation of motion is

$$v_x\frac{dv_x}{dx} = -\frac{1}{\rho}\frac{dp}{dx} \tag{6.17}$$

• **Bernoulli's equation**
Integration of (6.17) gives Bernoulli's equation

$$\frac{v_x^2}{2} + \int \frac{dp}{\rho} = E$$

This equation can be simplified for an isothermal or a polytropic flow by using (6.6) or (6.7)

$$\frac{v_x^2}{2} + v_{S,i}^2 \ln \frac{\rho(x)}{\rho(x_0)} = E$$

$$\frac{v_x^2}{2} + \frac{\alpha}{\alpha - 1} \frac{p}{\rho} = E \qquad (6.18)$$

E is a constant. A different value of E corresponds to a different flow. In (6.18) $v_x^2/2$ is the kinetic energy per unit mass and $v_{S,i}^2 \ln \frac{\rho(x)}{\rho(x_0)}$ (for the isothermal flow) or $(\alpha p)/((\alpha - 1)\rho)$ (for the polytropic flow) is the enthalpy per unit mass. Polytropic flow with $\alpha = \gamma$ is the standard approximation for flow in a nozzle. The results for isothermal flow are given for later comparison with Parker's isothermal solar wind theory.

• **Differential equation for v_x**
We now combine the equation of motion (6.17) and the continuity equation (6.16) and use the assumption that the flow is polytropic (adiabatic) to obtain the classic equation for v_x for flow in a nozzle

$$\frac{1}{v_x}\left(\frac{v_x^2}{v_S^2} - 1\right)\frac{dv_x}{dx} = -\frac{1}{\sigma}\frac{d\sigma}{dx} \qquad (6.19)$$

This is an ordinary non-linear first order differential equation for v_x. It is the second key equation of our discussion of the rocket engine. The first key equation is of course Bernoulli's equation. The solutions to the differential equation for v_x are given in implicit form by Bernoulli's equation. Note that in (6.19) v_S stands for the polytropic speed of sound. As it is agreed that $\alpha = \gamma$ is the best choice, v_S is the adiabatic speed of sound. The ratio v_x/v_S is the local Mach number M

$$M(x) = \frac{v_x}{v_S} \qquad (6.20)$$

It will play a decisive role in what follows. Let us have a closer look at the differential equation for v_x. The factor

$$\frac{v_x^2}{v_S^2} - 1 = M^2 - 1$$

plays a key role. This factor tells us whether the flow is *subsonic*: $M < 1$, *sonic*: $M = 1$ or *supersonic*: $M > 1$ and the behaviour of subsonic and supersonic flow is drastically different.

- Subsonic flow: $M < 1, \;\; v_x^2/v_S^2 < 1$
  The differential equation for v_x tells us that

$$\text{converging tube}: \frac{d\sigma}{dx} < 0 \quad \rightarrow \quad \frac{dv_x}{dx} > 0 : \text{flow accelerates}$$

$$\text{tube with constant cross-section}: \frac{d\sigma}{dx} = 0 \quad \rightarrow \quad \frac{dv_x}{dx} = 0 : \text{constant flow}$$

$$\text{diverging tube}: \frac{d\sigma}{dx} > 0 \quad \rightarrow \quad \frac{dv_x}{dx} < 0 : \text{flow decelerates}$$

This is similar to incompressible flow.

- Supersonic flow: $M > 1$, $v_x^2/v_{S,i}^2 > 1$
  The differential equation for v_x tells us that

$$\text{converging tube}: \frac{d\sigma}{dx} < 0 \quad \rightarrow \quad \frac{dv_x}{dx} < 0 : \text{flow decelerates}$$

$$\text{tube with constant cross-section}: \frac{d\sigma}{dx} = 0 \quad \rightarrow \quad \frac{dv_x}{dx} = 0 : \text{constant flow}$$

$$\text{diverging tube}: \frac{d\sigma}{dx} > 0 \quad \rightarrow \quad \frac{dv_x}{dx} > 0 : \text{flow accelerates}$$

Automotive traffic flow behaves as supersonic flow in a nozzle!

• Critical point and transonic solutions
The position x_s where

$$M = 1, \;\; v_x^2(x_s) = v_S^2(x_s) \tag{6.21}$$

is of particular importance for the nozzle problem. Actually we cannot avoid this point if we want to accelerate a flow from subsonic to supersonic. It is called a *sonic point* or a *critical point.* Since we exclude shocks and only consider continuous flows with finite derivatives of v_x, it follows that at a sonic point the right hand side of (6.19) must vanish. Sonic flow only occurs for $\frac{d\sigma}{dx} = 0$. For a tube with a non-constant cross-section $\sigma(x)$, the condition $M = 1$ or $v_x = v_S$ occurs at a point where $\sigma(x)$ attains an extremum. For the rocket engine this extremum is a minimum and the sonic point is located in the throat of the nozzle!

A rocket engine accelerates subsonic flow to supersonic flow with the obvious consequence that the flow passes through a sonic point. The same situation occurs in the solar wind. At the solar surface the flow is definitely subsonic and hardly observable, while at the earth's orbit the solar wind is supersonic. Thus, somewhere between the solar surface and the earth's orbit the solar wind passes through the sonic point. Let us design a rocket engine, i.e. a tube in which an initially subsonic flow is accelerated to supersonic speeds. We have all the information to do that!

- We start subsonic; accelerate the flow ($dv_x/dx > 0$) in a converging tube $d\sigma/dx < 0$.
- We become sonic $v_x = v_S$ at the sonic point. Since you have avoided shocks $d\sigma/dx = 0$ at $x = x_s$.
- We are now supersonic and accelerate supersonically ($dv_x/dx > 0$) in a diverging tube ($d\sigma/dx > 0$).

This sequence constitutes the basic principle of the de Laval nozzle or rocket engine and it is this profile $\sigma(x)$ that is shown on Fig. 6.4. So, we have one of the possible solutions of flow in a tube. Since it passes through the sonic point it is called a *transonic solution.*

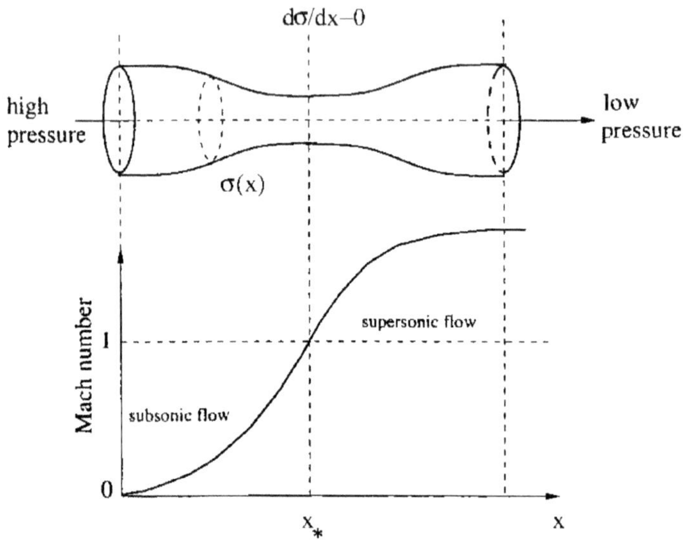

Figure 6.4: *A de Laval nozzle for accelerating a flow from subsonic to supersonic speeds.*

For comparison with Parker's isothermal solar wind model in the following Section, we go back to the isothermal version of (6.18). This equation can be used to express density in terms of velocity as

$$\rho(x) = \rho(x_0) \exp\left(-\frac{v_x^2 - v_x(x_0)^2}{2v_{S,i}^2}\right) \tag{6.22}$$

Finally recall that the basic physical principle for accelerating a flow in both the solar wind and the rocket engine is the conversion of enthalpy (random motions) in kinetic energy (directed motions). In the rocket engine, the chamber walls provide confinement; the critical point is located in the throat of the nozzle, and the supersonic expansion results. What are the analogs in the solar wind. Go to the following Section. The best is yet to come!

6.5 Parker's isothermal solution for a thermally driven wind

At the end of the third Section we came to the conclusion that

- static coronal models cannot be embedded in the interstellar medium;
- it is difficult to avoid expansion of a hot corona because of transfer of enthalpy to kinetic energy.

The previous Section has taught us how we can accelerate a subsonic flow to supersonic speeds in a rocket engine and we expect that the same acceleration mechanism works in the solar wind. Hence, this is a good time to take the next logical step and include a radial outflow in a simple equilibrium model. The model is defined as follows.

- It is in stationary equilibrium. It is time independent but it has a purely radial velocity field. Again we forget about the solar rotation.

- It is spherically symmetric. In a system of spherical polar coordinates (r, θ, φ) all quantities only depend on the radial distance to the centre of the sun. The only derivative that survives is d/dr.

- It is isothermal: $T = \text{constant}, v_{S,i} = \text{constant}$.

- There is no magnetic field.

Mass conservation and equation of motion

The outflow is steady, spherically symmetric and radial. The mass conservation equation (3.41) is

$$\frac{1}{r^2}\frac{d}{dr}\left(r^2 \rho v_r\right) = 0$$

This can be written as

$$\frac{1}{\rho}\frac{d\rho}{dr} = -\frac{2}{r} - \frac{1}{v_r}\frac{dv_r}{dr} \tag{6.23}$$

or integrated

$$r^2 \rho(r) v_r(r) = C_1 \tag{6.24}$$

where C_1 is a constant. The equation of motion (3.41) has only its radial component not identically equal to 0. As far as this equation is concerned, we have to modify the equation of hydrostatic equilibrium (6.8) by including the contribution due to the steady flow. Hence,

$$v_r \frac{dv_r}{dr} = -\frac{G\mathcal{M}_\odot}{r^2} - \frac{1}{\rho}\frac{dp}{dr} \tag{6.25}$$

Bernoulli's equation

Integration of (6.25) gives Bernoulli's equation

$$\frac{v_r^2}{2} - \frac{G\mathcal{M}_\odot}{r} + \int \frac{dp}{\rho} = E$$

This equation can be simplified for an isothermal or a polytropic flow by using (6.6) or (6.7)

$$\frac{v_r^2}{2} - \frac{G\mathcal{M}_\odot}{r} + v_{S,i}^2 \ln \frac{\rho(r)}{\rho(r_0)} = E$$

$$\frac{v_r^2}{2} - \frac{G\mathcal{M}_\odot}{r} + \frac{\alpha}{\alpha - 1}\frac{p}{\rho} = E \tag{6.26}$$

As in the nozzle problem (6.18) E is a constant and a different value of E corresponds to a different solution. In comparison to (6.18) there is one additional term in (6.26), $-G\mathcal{M}_\odot/r$, which is the gravitational potential energy per unit mass. As in the nozzle problem (6.26), $v_r^2/2$ is the kinetic energy per unit mass and $v_{S,i}^2 \ln \frac{\rho(r)}{\rho(r_0)}$ (for isothermal flow) or $\alpha p/((\alpha-1)\rho)$ is the enthalpy per unit mass.

Differential equation for v_r

We now combine the equation of motion (6.25) and the continuity equation (6.23) and use the assumption that the flow is either isotropic or polytropic to obtain the following differential equation for v_r

$$\frac{1}{v_r}\left(\frac{v_r^2}{v_S^2}-1\right)\frac{dv_r}{dr} = \frac{2}{r}\left(1-\frac{G\mathcal{M}_\odot/2v_S^2}{r}\right) \tag{6.27}$$

This is the classic Parker equation for v_r. It is an ordinary non-linear first order differential equation. It is instructive to compare (6.27) with the equation (6.19) for v_x in the nozzle problem. The left hand sides of these two equations are identical and the right hand sides show that the role of $(1/\sigma)(d\sigma/dx)$ for the nozzle is now played by $(2/r)(1 - G\mathcal{M}_\odot/(2v_{S,i}^2 r)$ for the radial outflow. The ratio v_r/v_S is the local Mach number M

$$M(r) = \frac{v_r}{v_S} \tag{6.28}$$

$G\mathcal{M}_\odot/2v_S^2$ has the dimension of length and we denote it as r_s

$$r_s = \frac{G\mathcal{M}_\odot}{2v_{S,i}^2} \tag{6.29}$$

For $T_0 = 1-2\times10^6$ K we find $r_s \approx 3-5R_\odot$ where the smaller/larger value of r_s corresponds to the higher/lower value of T_0. With the use of (6.29) the right hand side of (6.27) can be written in the following compact form

$$\text{RHS} = \frac{2}{r}(1-\frac{r_s}{r})$$

The index "s" indeed indicates that r_s is the location of the sonic point. A continuous flow with finite derivatives of v_r (dv_r/dr = finite) (i.e. a flow without shocks) can only be sonic at $r = r_s$:

$$M = 1, \ \ v_r^2/v_S^2 = 1 \ \ \Leftrightarrow \ \ r = r_s \tag{6.30}$$

In order to take the correspondence between the radial outflow and the flow in a nozzle further we note that

$$RHS < 0 \text{ for } r < r_s, \ \ RHS = 0 \text{ for } r = r_s, \ \ RHS > 0 \text{ for } r > r_s$$

The radial outflow is an exact copy of the rocket engine for the acceleration of subsonic flow at the solar surface to supersonic flow at the earth's orbit with the role of $(1/\sigma)d\sigma/dx$ now played by $2(1-r_s/r)/r$.

Transonic solutions

In the low corona (at $r = r_0$) the flow is definitely subsonic and hardly observable. The outflow velocities there are estimated to be of the order of (or smaller than) 10 km/s and are much smaller than the local sound velocity of about 200 km/s. On the other hand, at the earth's orbit the solar wind has an average velocity of about 400 km/s and is definitely supersonic. Thus, somewhere between the solar surface (low corona) and the earth's orbit the solar wind passes through the sonic point. Here is how nature does it!

- The flow starts subsonically at the solar surface. Since $2(1 - r_s/r)/r < 0$ for $r < r_s$ the flow accelerates $(dv_r/dr > 0)$ as if in a converging tube.
- It becomes sonic at the sonic point $r = r_s$ and becomes supersonic since $dv_r/dr > 0$.
- The flow is now supersonic and accelerates supersonically $(dv_r/dr > 0)$ since $2(1 - r_s/r)/r > 0$ for $r > r_s$ as if in a diverging tube.

This sequence constitutes the basic principle of the thermal acceleration of the solar wind. It is one of the mathematically possible solutions of radial outflow around the sun. Since it passes through the sonic point it is called a *transonic solution.* The clever reader has already figured out that there is a second transonic solution.

- The flow starts supersonically at the solar surface and decelerates $(dv_r/dr < 0)$ for $r < r_s$ as if in a converging tube.
- It becomes sonic at the sonic point $r = r_s$ and becomes subsonic since $dv_r/dr < 0$.
- The flow is now subsonic and decelerates subsonically $(dv_r/dr < 0)$ for $r > r_s$ as if in a diverging tube.

This solution is of no relevance for the solar wind.

Solutions as level curves of Bernoulli's function

The left hand side of Bernoulli's equation (6.26) contains the radial coordinate r and the three dependent variables ρ, p, v_r. Two of these three variables can be expressed in terms of the remaining third one. In order to make further progress we need to carry out this elimination for Bernoulli's equation. This is most easily done for an isothermal flow which is actually the case studied by Parker in his famous 1958 paper. The elimination process for polytropic flow is a little bit more involved but still very straightforward. For an isothermal flow we can eliminate the term $\ln(\rho(r)/\rho(r_0))$ by using (6.23). We prefer to relate velocity to $v_{S,i}$ and distance to r_s. This leads to the following second version of Bernoulli's equation for an isothermal flow

$$\boxed{\frac{v_r^2}{v_{S,i}^2} - \ln\frac{v_r^2}{v_{S,i}^2} - \ln\frac{r^4}{r_s^4} - 4\frac{r_s}{r} = C_* \qquad (6.31)}$$

Bernoulli's equation (6.31) is the solution of the differential equation (6.27) in implicit form. It tells us that that the solutions of (6.27) correspond to the level curves of the Bernoulli function $H(r, v_r)$

$$H(r, v_r) = \frac{v_r^2}{v_{S,i}^2} - \ln\frac{v_r^2}{v_{S,i}^2} - \ln\frac{r^4}{r_s^4} - 4\frac{r_s}{r} \qquad (6.32)$$

The solutions correspond to the level curves of the function $H(r, v_r)$. A different value of the constant C_* defines a different level curve and thus a different solution. Of course there is an infinite number of level curves of $H(r, v_r)$ and thus an infinite number of solutions for the radial outflow from the sun.

Transonic solutions again

We focus on the transonic solution and determine the value of C_* that corresponds to the transonic solution. At the sonic point $r = r_s = G\mathcal{M}_\odot/2v_{S,i}^2$ we have $v_r = v_{S,i}$ so that $C_{*,tr} = -3$. The transonic solution is then

$$\frac{v_r^2}{v_{S,i}^2} - \ln\frac{v_r^2}{v_{S,i}^2} = -3 - \ln\frac{r_s^4}{r^4} + 4\frac{r_s}{r} \tag{6.33}$$

We start at $r = r_0$ and assume that $r_0 < r_s$. (Is this always true? What happens if this inequality is not fulfilled?) What velocity $v_r(r_0)$ do we need there to get a ride on the transonic solution? Obviously $v_r(r_0)$ has to satisfy (6.33) with r replaced with r_0. This is a transcendental equation for $v_r(r_0)/v_{S,i} = M_0$. It is straightforward to show that $\forall x > 1 :$ $x - \ln x > 1$, so that under the assumption that $r_0 < r_s$

$$-3 - 4\ln\frac{r_s}{r_0} + 4\frac{r_s}{r_0} = \alpha > 1$$

The equation for M_0 is then

$$M_0^2 - \ln M_0^2 = \alpha$$

and has two roots $M_{0,1}$ and $M_{0,2}$ with

$$M_{0,1} \in]0,\ 1[\ \wedge\ M_{0,2} > 1$$

Hence, we have two transonic solutions. The one corresponding to $M_{0,1}$ has already been discussed extensively. The transonic solution corresponding to $M_{0,2}$ starts supersonically, decelerates, goes through the sonic point and further decelerates subsonically. It is of no interest for the solar wind as supersonic outflows are not present in the low corona. An accurate approximation of $M_{0,1}$ can be obtained as follows. With an iterative scheme we can obtain the solution with prescribed accuracy. However, it is instructive from the viewpoint of physics and helpful from the viewpoint of numerical mathematics to obtain an approximate solution. We have to remember that the solution is strongly subsonic at $r = r_0$. Since

$$v_r(r_0) << v_{S,i}, \quad \frac{v_r(r_0)^2}{v_{S,i}^2} << -\ln\left(\frac{v_r(r_0)^2}{v_{S,i}^2}\right)$$

the equation (6.33) for $v_r(r_0)$ can be simplified to

$$\begin{aligned} -2\ln\left(\frac{v_r(r_0)}{v_{S,i}}\right) &\approx -3 - 4\ln\frac{r_s}{r_0} + 4\frac{r_s}{r_0} \\ \ln\left(\frac{v_r(r_0)}{v_{S,i}}\right) &\approx \ln\left\{(r_s/r_0)^2\exp(3/2 - 2r_s/r_0)\right\} \end{aligned}$$

The solution to this approximate equation is

$$\frac{v_r(r_0)}{v_{S,i}} \approx (r_s/r_0)^2 \exp(3/2 - 2r_s/r_0)$$

This isothermal prediction for the outflow velocity in the low corona can be used to get qualitative information. The predicted values are of the order 1 – 10 km/s depending on the value used for T_0.

Other solutions

The differential equation (6.27) for v_r has an infinite number of solutions corresponding to the level curves of the Bernoulli function (6.32). In order to understand the level curves of the Bernoulli function, we need to determine the critical points of (6.32) and to see whether these critical points correspond to an extremum or a saddle point. We compute the first order partial derivatives $\partial H/\partial r$ and $\partial H/\partial v_r$ and find that there is just one critical point

$$r = r_s, \quad v_r = v_{S,i}$$

which is the sonic point (6.30). The next step is to find out whether this critical point corresponds to an extremum (that would be bad, why?) or to a saddle point of $H(r, v_r)$! Hence, we determine the second order partial derivatives $\partial^2 H/\partial r^2$, $\partial^2 H/\partial r \partial v_r$ and $\partial^2 H/\partial v_r^2$. We compute their values at the critical point and find that the critical point is a saddle point of the function $H(r, v_r)$. This is very good news. The level curves of $H(r, v_r)$ do not encircle the saddle point. There are only two level curves that pass through the sonic point; i.e. the two transonic solutions discussed earlier. The two transonic solutions divide the (r, v_r) plane in four regions, which are labelled I to IV as shown on Fig. 6.5. This figure is borrowed from C1994 "Large-scale Dynamics of the Solar Wind", unpublished class notes, 1994. The isothermal sound speed is shown as a dotted line. In C1994's notation the isothermal sound speed is denoted as a. Region I contains the solutions that start supersonically and remain supersonically all the time. Conversely, the solutions in region III start subsonically and remain subsonically all the time. The solutions in region II are confined to $r > r_s$ and disconnected from the sun; the solutions in region IV are confined to $r < r_s$ and do not connect to the interplanetary space. The only physically relevant solution is the transonic solution with $dv_r/dr > 0$. It is the only solution that starts with subsonic flow at the solar surface and expands in a supersonic flow at the earth's orbit and beyond! This is our favorite solution. The solutions in region III are referred to as the solar breeze. The observations by satellites of the velocities of the solar wind convinced people that the solar breeze is not physically relevant!

Density-velocity relation and mass loss

The isothermal version of Bernoulli's equation (6.26) can be used to determine the variation of density with distance r. The constant E in (6.26) can be readily determined in terms of the values at the position r_0 as

$$E = \frac{v_r^2(r_0)}{2} - \frac{G\mathcal{M}_\odot}{r_0}$$

We can combine (6.26) and the expression for E to obtain the density velocity relation.

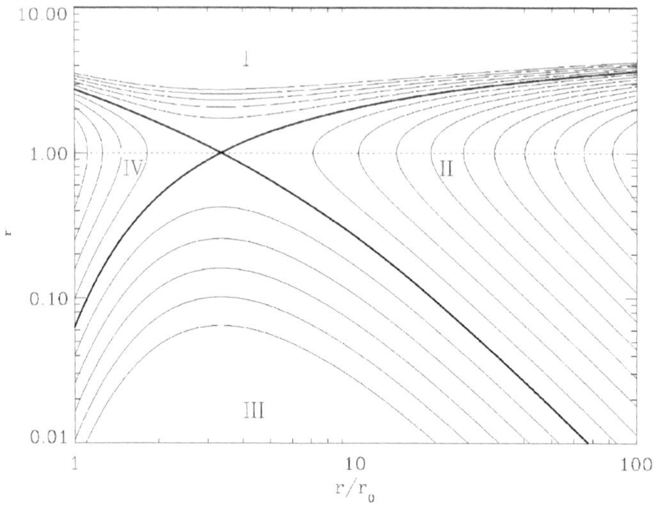

Figure 6.5: *Topology of the isothermal Parker solutions for a thermally driven wind. Each curve is a level curve of the Bernoulli function and corresponds to a different value of E. The two transonic solutions divide the* (r, v_r) *plane in four regions. (C1994).*

$$\rho(r) = \rho_0 \exp(\underbrace{-\frac{GM_\odot/v_{S,i}^2}{r_0}(1-\frac{r_0}{r})}_{\text{static / isothermal}} \underbrace{-\frac{v_r^2 - v_r^2(r_0)}{2v_{S,i}^2}}_{\text{outflow}}) \qquad (6.34)$$

It is instructive to compare expression (6.34) with that obtained for the static isothermal atmosphere stratified by gravity (6.10) and with that obtained for the nozzle (6.22). Remarkable isn't it? It is straightforward to verify that the (absolute value of the) ratio of the second to the first term in the exponential increases with distance. At the sonic point this ratio is $\approx 1/(4(r_s - r_0)) \approx 0.1$, so that below $r = r_s$ the structure of the corona is very much like that of a static isothermal corona. Beyond r_s the ratio increases rapidly, meaning that the expansion takes over and that the structure is markedly different from that of a static corona. Since $v_r(r_0) << v_{S,i}$ we can obtain from (6.34) that

$$\ln\frac{\rho(r_s)}{\rho(r_0)} \approx \frac{3}{2} - 2\frac{r_s}{r_0} \qquad (6.35)$$

This allows us to compute $\rho(r_s)$ in terms of ρ_0.

The mass that the sun loses per unit time due to the wind is

$$\frac{dM}{dt} = 4\pi r_0^2 \rho(r_0) v_r(r_0) = 4\pi r_s^2 \rho(r_s) v_S \qquad (6.36)$$

For an isothermal wind we can use the definition (6.29) of r_s and rewrite (6.36) as

$$\frac{dM}{dt} = \frac{\pi(G\mathcal{M}_\odot)^2}{v_{S,i}^2}\rho(r_s) \tag{6.37}$$

$\rho(r_s)$ can be expressed in terms of $\rho(r_0)$ (6.35). Hence, we can compute dM/dt if we can agree on the best values for r_0 and ρ_0. Anyway, it is not too difficult by shuffling the numbers to get a theoretical mass loss rate of the order of $10^{-14}\mathcal{M}_\odot/\mathrm{yr}$ which is remarkably close to what is observed. We should not overrate this result as it is obtained under the assumption that the outflow is isothermal.

6.6 Rotating thermally driven wind

Parker's isothermal radial solar wind model does a remarkably good job in reproducing the observed velocities and densities at the earth's orbit. By now the interested reader should have discovered in Problem 13 that the velocity of the wind diverges at large heliospheric distances, This behaviour is due to the assumption that the wind is isothermal. This assumption implies that energy is pumped into the wind all the way up to infinity and that the wind keeps on accelerating. When the assumption that the wind is isothermal is replaced by the assumption that it is polytropic the wind will have a finite velocity at infinity. All stars and in particular the sun rotate and it is not a priori clear that rotation is unimportant for stellar winds. An obvious reason why we cannot forget about rotation when constructing a realistic wind model is that a stellar wind carries angular momentum away from the star and helps to brake the rotation of the star. Braking mechanisms for stellar rotation are in high demand since pre-main sequence contraction and main sequence evolution without an effective braking mechanism would make all stars to be very fast rotators and that is not what is observed. The sun is a slow rotator as can be seen from the ratio of its rotational kinetic energy to its gravitational potential energy. So we can anticipate that the effect of rotation on the solar wind is weak.

Let us take the next logical step and include rotation in the wind model. Rotation complicates life seriously since spherical symmetry no longer holds. This is because the centrifugal force does not point away or toward the origin, but rather is perpendicular to the rotation axis. Axial symmetry with the rotation axis defining the symmetry axis is preserved. For simplicity, solutions are constructed in the equatorial plane. As in the previous Section the solutions are time independent and nonmagnetic! Let us recapitulate: the model has the following properties.

- It is in stationary equilibrium. It is time independent so that $\frac{\partial}{\partial t} = 0$.
- It is axially symmetric with the rotation axis as symmetry axis. In a system of spherical polar coordinates (r, θ, φ) with the z-axis coinciding with the rotation axis all quantities are independent of the angle φ or $\frac{\partial}{\partial \varphi} = 0$.
- The model is confined to the equatorial plane. This implies that $\frac{\partial}{\partial \theta} = 0$ and $v_\theta = 0$.
- It is nonmagnetic.

Axial symmetry and the restriction to solutions in the equatorial plane imply that

$$\vec{v} = (v_r(r), 0, v_\varphi(r)) \tag{6.38}$$

and also that the only derivative that survives is d/dr. At the reference position r_0

$$v_\varphi(r) = r_0 \Omega_\odot$$

where $\Omega_\odot$ is the angular velocity of the sun.

Mass conservation and equation of motion

The equation for mass conservation remains unchanged. We can borrow the results (6.23) and (6.24) from the previous Section. As far as the equation of motion is concerned life becomes more complicated. The radial and axial components of the equation of motion are respectively

$$\begin{aligned} v_r \frac{dv_r}{dr} - \frac{v_\varphi^2}{r} &= -\frac{G\mathcal{M}_\odot}{r^2} - \frac{1}{\rho}\frac{dp}{dr} \\ v_r \frac{dv_\varphi}{dr} + \frac{v_r v_\varphi}{r} &= 0 \end{aligned} \tag{6.39}$$

The radial component of the equation of motion was already present in the previous Section and integration of it gave us Bernoulli's equation. This will also be the case here. The new information is contained in the axial component. This component will tell us about angular momentum.

Angular momentum

From the axial component of the equation of motion (6.39) follows that angular momentum per unit mass, rv_φ, is constant

$$rv_\varphi = L$$

where L is a constant, which is readily determined as $L = r_0 v_\varphi(r_0) = r_0^2 \Omega_\odot$ so that

$$rv_\varphi = L = r_0^2 \Omega_\odot \tag{6.40}$$

Hence

$$v_\varphi = \frac{L}{r} = \frac{r_0^2 \Omega_\odot}{r} \tag{6.41}$$

In a non-radial wind of a rotating star the azimuthal component of velocity decreases with distance as $1/r$.

Bernoulli's equation

We use (6.41) to rewrite the equation of motion as

$$v_r \frac{dv_r}{dr} - \frac{r_0^4 \Omega_\odot^2}{r^3} + \frac{G\mathcal{M}_\odot}{r^2} + \frac{1}{\rho}\frac{dp}{dr} = 0 \tag{6.42}$$

Integration of this equation gives Bernoulli's equation

$$\frac{v_r^2 + v_\varphi^2}{2} - \frac{G\mathcal{M}_\odot}{r} + \int \frac{dp}{\rho} = E$$

This equation can be simplified for an isothermal or a polytropic flow by using (6.6) or (6.7)

$$\frac{v_r^2 + v_\varphi^2}{2} - \frac{G\mathcal{M}_\odot}{r} + v_{S,i}^2 \ln \frac{\rho(r)}{\rho(r_0)} = E$$

$$\frac{v_r^2 + v_\varphi^2}{2} - \frac{G\mathcal{M}_\odot}{r} + \frac{\alpha}{\alpha - 1} \frac{p}{\rho} = E \qquad (6.43)$$

We can compare (6.43) with its non-rotational counterpart (6.26). There is one additional term in (6.43), i.e. $v_\varphi^2/2$. It is the kinetic energy per unit mass due to rotation.

Differential equation for v_r

We now use the equation of mass conservation and the assumption that the flow is isothermal or polytropic to express the fourth term on the left hand side of (6.42) in terms of v_r and r. With little additional rearranging we obtain the differential equation for v_r

$$\frac{1}{v_r}\left(\frac{v_r^2}{v_S^2} - 1\right)\frac{dv_r}{dr} = \frac{2}{r}\left(1 - \frac{G\mathcal{M}_\odot/2v_S^2}{r} + \underbrace{\frac{\tau^2}{2}\frac{r_0^2}{r^2}}\right) \qquad (6.44)$$

(6.44) is the modification of Parker's equation for v_r (6.27) when rotation is present. As in (6.27) v_S is either the isothermal or the polytropic velocity of sound. Comparison of equation (6.44) with (6.27) for the non-rotating wind model shows that there is one additional term on the right hand side. It is underbraced for clarity. The quantity τ is defined as

$$\tau = \frac{r_0 \Omega_\odot}{v_S} = \frac{v_\varphi(r_0)}{v_S}$$

and measures the relative importance of rotation. The sun rotates differentially. Its equatorial rotational period is 26 days. The corresponding rotational frequency is $\Omega_\odot = 2.7970 \times 10^{-6}$ rad/s. The rotational velocity at the surface of the sun is then $R_\odot \Omega_\odot \approx 2 \times 10^3$ m/s. For $r_0 = 1.2R_\odot$ and $T = 1 - 2 \times 10^6$ K we find $\tau_\odot \approx 0.01$. This is a strong indication that rotation will not introduce drastic changes in the solar wind theory.

What about the critical points? The plural is right; there can be two of them, or none or one! The position(s) $r = r_s$ where a continuous flow with finite derivatives of v_r (dv_r/dr = finite) (i.e. a flow without shocks) can be sonic

$$M = 1, \ \ v_r^2/v_S^2 = 1 \ \Leftrightarrow \ r = r_s$$

is (are) determined by the equation

$$1 - \frac{G\mathcal{M}_\odot/2v_{S,i}^2}{r} + \frac{\tau^2}{2}\frac{r_0^2}{r^2} = 0$$

which we can rewrite as a quadratic equation in r as

$$r^2 - r_{s,0} r + \frac{\tau^2 r_0^2}{2} = 0 \tag{6.45}$$

$r_{s,0}$ denotes the position of the critical point in absence of rotation as the attentive reader had already figured out. It is given by (6.29). The number of critical points depends on the rotational velocity. For the case that there is a critical point in absence of rotation, i.e. $r_{s,0} > r_0$ we (actually the reader) can define two critical rotational velocities

$$\tau_{c,1} = \sqrt{2}\left(\frac{G\mathcal{M}_\odot}{2v_S^2 r_0} - 1\right)^{1/2}, \quad \tau_{c,2} = \frac{G\mathcal{M}_\odot}{2\sqrt{2}v_S^2 r_0}$$

so that

$$\begin{aligned} 0 \le \tau < \tau_{c,1} &\quad : \quad \text{1 critical point} \\ \tau_{c,1} \le \tau < \tau_{c,2} &\quad : \quad \text{2 critical points} \\ \tau = \tau_{c,2} &\quad : \quad \text{1 critical point} \\ \tau > \tau_{c,2} &\quad : \quad \text{no critical point} \end{aligned}$$

The sun with its small value of τ has one critical point as in the non-rotating case.

Level curves of Bernoulli's function

As in the non-rotating case (6.26) the left hand side of Bernoulli's equation (6.43) contains the radial coordinate r and the three dependent variables ρ, p, v_r. Two of these three variables can be expressed in terms of the remaining third one. In order to make further progress we need to carry out this elimination for Bernoulli's equation. Again, as in the non-rotating case, this is most easily done for an isothermal flow. The elimination process for polytropic flow is a little bit more involved but still very straightforward. For isothermal flow we can eliminate the term $\ln \frac{\rho(r)}{\rho(r_0)}$ by using (6.24). We prefer to relate velocity to $v_{S,i}$ and distance r_s. This leads to the following second version of Bernoulli's equation for an isothermal flow

$$\boxed{\frac{v_r^2}{v_{S,i}^2} - \ln\frac{v_r^2}{v_{S,i}^2} - \ln\frac{r^4}{r_s^4} - \frac{2G\mathcal{M}_\odot/v_{S,i}^2}{r} + \underbrace{\tau^2\frac{r_0^2}{r^2}} = C_* \qquad (6.46)}$$

It is instructive to compare (6.46) with its non-rotating counterpart (6.31). The rotational version of Bernoulli's equation has one additional term, which is underbraced. Bernoulli's equation is the solution of the differential equation in implicit form. Hence it is logical to define the Bernoulli function $H(r, v_r)$ as

$$H(r, v_r) = \frac{v_r^2}{v_{S,i}^2} - \ln\frac{v_r^2}{v_{S,i}^2} - \ln\frac{r^4}{r_s^4} - \frac{2G\mathcal{M}_\odot/v_{S,i}^2}{r} + \underbrace{\tau^2\frac{r_0^2}{r^2}} \tag{6.47}$$

The solutions of (6.44) correspond to the level curves of the function $H(r, v_r)$. A different value of the constant C_* defines a different level curve and thus a different solution. As before in the non-rotating case there is an infinite number of level curves of $H(r, v_r)$ and thus an infinite number of solutions for the outflow from the sun.

Transonic solutions again

We focus on the transonic solution and determine the value of C_* that corresponds to the transonic solution. At the critical point(s) $r = r_s$ we have $v_r = v_S$ and r_s is a solution of (6.45) so that $C_{*,tr} = -3 - \tau^2 r_0^2/r_s^2$. The transonic solution is then

$$\frac{v_r^2}{v_{S,i}^2} - \ln\frac{v_r^2}{v_{S,i}^2} = -3 - \ln\frac{r_s^4}{r^4} + 4\frac{r_s}{r} - \underbrace{\frac{\tau^2 r_0^2}{r_s^2}\left(\frac{r_s}{r} - 1\right)^2} \tag{6.48}$$

We start at $r = r_0$ and we assume that the wind has indeed a throat so that $r_0 < r_s$, which is definitely the case for the sun. Then we want to know what velocity $v_r(r_0)$ we need there to ride on the transonic solution. Obviously $v_r(r_0)$ has to satisfy (6.48) with r replaced with r_0. This is a transcendental equation for $v_r(r_0)/v_{S,i} = M_0$. In the previous Section we could show that the right hand of (6.33) is strictly larger than 1. However, this result is not any longer guaranteed for any value of τ. In case of the sun the right hand side of (6.48) for $r = r_0$ is still strictly larger than 1 because of the small solar value of τ. So for the sun the equation for M_0 has two roots $M_{0,1}$ and $M_{0,2}$ with

$$M_{0,1} \in]0,\ 1[\ \wedge\ M_{0,2} > 1$$

and there are still two transonic solutions. The one corresponding to $M_{0,1}$ has already been discussed extensively. The second transonic solution corresponding to $M_{0,2}$ starts supersonically, decelerates, goes through the sonic point and further decelerates subsonically. As before, it is of no interest for the solar wind.

Other solutions

As in the non-rotating case (6.27), equation (6.44) has an infinite number of solutions corresponding to the level curves of the function $H(r, v_r)$. We have to go through exactly the same routine as in the previous Section when discussing the solutions for the non-rotating wind. First we compute the first order partial derivatives $\partial H/\partial r$ and $\partial H/\partial v_r$ and find that the critical points $r = r_s,\ v_r = v_{S,i}$ are the sonic points (remember there can be 1, 2 or not any all depending on the value of τ!). For $\tau_\odot$ there is only one critical point and we can repeat the analysis of the previous Section to show that it corresponds to a saddle point of $H(r, v_r)$. Hence, we can repeat the arguments of the previous Section and conclude that the transonic solution with $dv_r/dr > 0$ is the only physically relevant solution, just as in the non-rotating case.

Density-velocity relation and mass loss

The isothermal version of Bernoulli's equation (6.43) can be used to determine the variation of density with distance r. The constant E in (6.43) can be determined in terms of the values at the position r_0 as

$$E = \frac{v_r^2(r_0) + v_\varphi^2(r_0)}{2} - \frac{G\mathcal{M}_\odot}{r_0}$$

We can combine (6.43) and the expression for E to obtain the density velocity relation

$$\rho(r) = \rho_0 \exp(\underbrace{-\frac{G\mathcal{M}_\odot / v_{S,i}^2}{r_0}(1 - \frac{r_0}{r})}_{\text{static / isothermal}} \underbrace{-\frac{v_r^2 - v_r^2(r_0)}{2v_{S,i}^2}}_{\text{outflow}} \underbrace{-\frac{\tau^2}{2}(1 - \frac{r_0^2}{r^2})}_{\text{rotation}}) \tag{6.49}$$

This result (6.49) for the variation of density with distance from the solar or stellar surface can be compared with that obtained for (i) the static isothermal atmosphere stratified by gravity (6.10), (ii) the nozzle (6.22) and (iii) the radial Parker outflow (6.34). The effect due to rotation is always small for $\tau = \tau_\odot$ and the difference between the rotational solution and Parker's solution is always small for the sun. For the same reason (6.35) remains valid for the rotating solar wind.

The mass that the sun or a star loses per unit time due to the wind is given by (6.36). For a slowly rotating star, such as the sun, and an isothermal wind we can use the definition (6.29) of r_s and rewrite the expression for mass loss as (6.37). For the sun, rotation is too slow to have any significant impact on the wind. For fast rotating stars with $\tau \approx 1$ the story can be drastically different from that for their non-rotating twins.

6.7 Rotating magnetized thermally driven wind

In the previous Section we have seen that rotation does not cause any substantial changes in the solar wind. In particular the transport of angular momentum by the solar wind is small and the non-magnetized version of the solar wind is not an effective braking mechanism. Let us now try and find out what the effect of the solar magnetic field on the solar wind is. Naively we might anticipate that we can forget about the solar magnetic field altogether. The global solar magnetic field is very weak: its energy is several orders of magnitude smaller than the sun's gravitational potential energy and in addition it is even much smaller than the sun's rotational energy. The sun is to a high degree of accuracy a spherically symmetric star that is held together by its own gravity and its small oblateness is due to its rotation. Nevertheless, we cannot forget about the solar magnetic field. The sun has a hot corona and this corona owes its existence to the magnetic field and without the hot corona there would not be any solar wind. But does the solar magnetic field affect the wind once the wind is there?

Let us take the next logical step and include a magnetic field in the rotational wind model. Weber and Davis (Weber, E.J. and Davis, L. Jr. : 1967, Ap.J. **148**, 217) were the first to construct a model for a magnetic rotating solar wind. It is 2-dimensional, i.e. constrained to the equatorial plane and polytropic. Magnetic fields complicate life even more than rotation. But this does not bother us as it means more fun; what we do not want at all is a dull life! Spherical symmetry no longer holds. This is because the Lorentz force and the centrifugal force are not central forces. Axial symmetry is preserved if the axis of symmetry of the magnetic field coincides with the rotation axis. For simplicity, solutions are constructed in the equatorial plane. As in the previous Sections the solutions are time independent! Let us recapitulate: the model has the following properties.

- It is in stationary equilibrium. It is time independent so that $\frac{\partial}{\partial t} = 0$.
- It is axially symmetric with the rotation axis = magnetic axis as symmetry axis: in a system of spherical polar coordinates (r, θ, φ) with the z-axis coinciding with the rotation axis all quantities are independent of the angle φ or $\frac{\partial}{\partial \varphi} = 0$.

- It is confined to the equatorial plane. Hence, $\frac{\partial}{\partial \theta} = 0$ and $v_\theta = 0$.
- The magnetic field is symmetric with respect to the equatorial plane: $B_\theta = 0$.

Axial symmetry and the restriction to solutions in the equatorial plane imply that

$$\vec{v} = (v_r(r), 0, v_\varphi(r)), \quad \vec{B} = (B_r(r), 0, B_\varphi(r)) \text{ for } \theta = \frac{\pi}{2} \tag{6.50}$$

and also that the only derivative that survives is d/dr. The angular velocity $\Omega_\odot$, density ρ, temperature T and the radial component of the magnetic field B_r are known at the reference position r_0:

$$\rho(r_0) = \rho_0, \quad T(r_0) = T_0, \quad v_\varphi(r_0) = r_0 \Omega_\odot, \quad B_{r,0} = B_0, \quad \text{known.}$$

What about $B_{\varphi,0}$?

Kinematics

Mass conservation

From (6.24) we know that $r^2\rho(r)v_r(r) = C_1$ where C_1 is a constant.

No magnetic monopoles

Since all quantities, by assumption, only depend on the radial coordinate r, the Maxwell equation $\nabla \cdot \vec{B} = 0$, reduces to

$$\frac{1}{r^2}\frac{d}{dr}(r^2 B_r) = 0$$

and tells us that $r^2 B_r = C_2$. C_2 is a constant which we can express in terms of values at r_0 as $C_2 = r_0^2 B_0$ so that

$$\boxed{B_r = B_0 \frac{r_0^2}{r^2} \qquad (6.51)}$$

The radial component of the magnetic field decreases as $1/r^2$ with distance from the sun. We can combine (6.24) and (6.51) to find

$$\frac{B_r}{\rho v_r} = \frac{C_2}{C_1} = \text{ constant}, \quad \frac{d}{dr}\left(\frac{B_r}{\rho v_r}\right) = 0 \tag{6.52}$$

Induction equation and Archimedean spirals

In view of the very high values of the Lundquist number for the solar wind plasma the ideal MHD version of the induction equation can be used. The steady state version is

$$\nabla \times (\vec{v} \times \vec{B}) = 0 \tag{6.53}$$

Straightforward calculation shows that for the outflow and the magnetic field considered here (6.50)

$$\nabla \times (\vec{v} \times \vec{B}) = \frac{1}{r}\frac{d}{dr}\left\{r(\vec{v} \times \vec{B})_\theta\right\} \vec{1}_\varphi$$

so that (6.53) leads to

$$v_\varphi B_r - v_r B_\varphi = \frac{C_3}{r}$$

C_3 is a constant which we can determine by using values at r_0:

$$\underbrace{v_\varphi(r_0)}_{=\Omega_\odot r_0}\underbrace{B_r(r_0)}_{=B_0}-v_r(r_0)\underbrace{B_\varphi(r_0)}_{\approx 0}=\frac{C_3}{r_0}$$

so that

$$C_3=\Omega_\odot r_0^2 B_0,\quad \frac{C_3}{r}=\frac{\Omega_\odot r_0^2 B_0}{r}=\Omega_\odot r B_r$$

Hence the induction equation leads to

$$B_r(v_\varphi - r\Omega_\odot) - v_r B_\varphi = 0 \qquad (6.54)$$

The relation (6.54) between $v_r, v_\varphi, B_r, B_\varphi$ is a direct consequence of flux freezing. At large distances from the sun $r >> r_s$ (e.g. at the earth's orbit) the outflow is practically radial: $v_\varphi \approx 0,\ \ v_r \approx V =$ constant and (6.54) takes the simple form

$$\frac{B_r}{B_\varphi}=\frac{-V}{r\Omega_\odot}$$

The equation of the magnetic field lines in the plane $\theta = \pi/2$ is

$$\frac{1}{r}\frac{dr}{d\varphi}=\frac{B_r}{B_\varphi}$$

Combination of the two previous equations means that at large distances from the sun

$$\frac{dr}{d\varphi}=-\frac{V}{\Omega_\odot}$$

The solution to this simple equation is

$$r = r_* - \frac{V}{\Omega_\odot}(\varphi - \varphi_*) \qquad (6.55)$$

and represents an Archimedean spiral (E. N. Parker, 1963, Interplanetary Dynamicall Proceses, Wiley). Due to the rotation of the sun the field lines of the solar magnetic field that is advected outward by the solar wind are Archimedean spirals as is shown on Fig. 6.6. This figure is borrowed from C1994 "Large-scale Dynamics of the Solar Wind", unpublished class notes, 1994. We denote the angle that the magnetic field vector makes with the radial direction as Ψ so that

$$B_r = B\cos\Psi,\quad B_\varphi = -B\sin\Psi$$

At the Earth's orbit the value of the angle Ψ predicted by this rotating magnetized wind model satisfies

$$\tan\Psi_E = -\left(\frac{B_\varphi}{B_r}\right)_E \approx \frac{r_E\Omega_\odot}{V}$$

Using the values $r_E = 1.50\times10^{11}$ m $= 215R_\odot$, $\Omega_\odot = 2.8\times10^{-6}$ rad/s and $V \approx 400$ km/s we get

$$\tan\Psi_E \approx 1,\quad \Psi_E \approx \frac{\pi}{4}$$

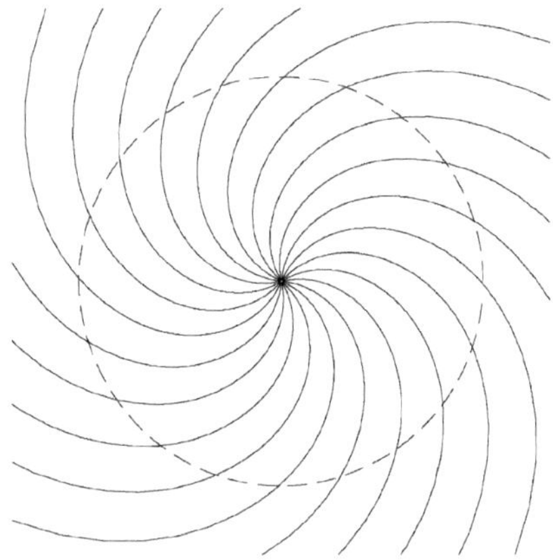

Figure 6.6: *The Archimedean spirals drawn by the magnetic field lines as they are advected outward by the solar wind. Solid lines correspond to magnetic field lines, and the circular dashed line corresponds to the Earth's orbit. The wind itself flows radially outward from the Sun. The magnetic field vector makes an angle* Ψ *with the radial direction.* B_φ *and* v_φ *have opposite signs.*

This is in remarkably good agreement with the observations. The measurements of the direction of the magnetic field vector show spatial and temporal variations but on average $\Psi_E \approx \pi/4$. This is seen as experimental confirmation of Parker's theory for the solar wind variation of the heliospheric magnetic field.

Dynamics

The radial and axial components of the equation of motion are respectively

$$\begin{aligned} \rho\left(v_r\frac{dv_r}{dr} - \frac{v_\varphi^2}{r}\right) &= -\frac{dp}{dr} - \rho\frac{G\mathcal{M}_\odot}{r^2} - \frac{B_\varphi}{\mu r}\frac{d}{dr}\left(rB_\varphi\right) \\ \rho\left(v_r\frac{dv_\varphi}{dr} + \frac{v_r v_\varphi}{r}\right) &= \frac{B_r}{\mu r}\frac{d}{dr}\left(rB_\varphi\right)) \end{aligned} \tag{6.56}$$

At this point it is not only politically correct but pedagogically sound to draw the reader's attention to the following two points.

- In the classic non-rotating Parker wind there is only a radial component of the equation of motion and manipulation of that radial component gave us Bernoulli's equation.
- In the rotating Parker wind the equation of motion has both a radial and axial component: the radial component leads to Bernoulli's equation and the axial component is related to conservation of angular momentum.

Angular momentum

In the rotating non-magnetized wind we have seen that the angular momentum per unit mass, rv_φ, is constant (6.40). However in a magnetic wind there is a torque associated with the magnetic field in addition to the angular momentum associated with the moving plasma element rv_φ. The φ-component of the equation of motion (6.56) can be manipulated into

$$\frac{d}{dr}\left(rv_\varphi - \frac{rB_\varphi B_r}{\mu\rho v_r}\right) = 0$$

which means that

$$rv_\varphi - \frac{rB_\varphi B_r}{\mu\rho v_r} = L \qquad (6.57)$$

with L the constant angular momentum per unit mass, rv_φ the angular momentum associated with the moving plasma element per unit mass and $(-rB_\varphi B_r)/(\mu\rho v_r)$ the torque per unit mass associated with the magnetic field. It is now convenient to introduce the Alfvén velocities $v_{A,r}$ and $v_{A,\varphi}$ associated with the radial and azimuthal components of the magnetic field as

$$v_{A,r}^2 = \frac{B_r^2}{\mu\rho}, \quad v_{A,\varphi}^2 = \frac{B_\varphi^2}{\mu\rho} \qquad (6.58)$$

Of course

$$v_A^2 = v_{A,r}^2 + v_{A,\varphi}^2$$

We now use (6.54) to eliminate $B_\varphi = (B_r(v_\varphi - r\Omega))/v_r$ to rewrite the equation for the conservation of angular momentum (6.57) as

$$v_\varphi = \frac{r\Omega_\odot}{M_A^2 - 1}\left(L\frac{M_A^2}{r^2\Omega} - 1\right) \qquad (6.59)$$

M_A is the Alfvén Mach number. It is defined as

$$M_A = \frac{v_r(r)}{v_{A,r}(r)} \qquad (6.60)$$

$v_{A,r}(r)$, $v_r(r)$ and $M_A(r)$ are functions of r. In the low corona the local Alfvén velocity is of the order $10^2 - 10^3$ km/s while the outflow velocity is of the order or less than 10 km/sec. Hence the solar wind is definitely sub-Alfvénic in the low corona. On the other hand the average Alfvén velocity and the average solar wind velocity at the Earth's orbit are 50 - 60 km/s and 400 - 500 km/s respectively. At the Earth's orbit the solar wind is super-Alfvénic. Hence, somewhere between the solar surface (low corona) and the earth's orbit the solar wind passes through a point r_A, where

$$M_A(r_A) = 1, \quad v_r(r_A) = v_{A,r}(r_A) \qquad (6.61)$$

This distance r_A is the Alfvén radius and defines the Alfvénic point or better the Alfvénic sphere where the energy densities of the flow and the magnetic field are equal. At the Alfvén point the solar wind changes from sub-Alfvénic to super-Alfvénic. At the Alfvénic point the denominator of the right hand side of (6.59) vanishes. Hence, in order to have a finite value for v_φ at the Alfvén point $r = r_A$ it is required that

$$L\frac{v_r(r_A)^2}{r_A^2\Omega_\odot} - v_{A,r}^2(r_A) = 0$$

so that

$$\boxed{L = r_A^2\Omega_\odot} \qquad (6.62)$$

This is a crucial result. It should be compared with (6.40). The important point to note is that r_0 is replaced with r_A. Hence, (6.62) tells us that the total angular momentum carried away by the magnetized rotating wind is equal to the angular momentum carried away by an non-magnetized wind flowing strictly radially and co-rotating out to a radius r_A. r_A is the effective lever arm used by the wind to brake the Sun's rotation. We can use (6.62) to rewrite the expression for v_φ as

$$v_\varphi = \frac{r\Omega_\odot}{M_A^2 - 1}\left(\frac{M_A^2}{(r/r_A)^2} - 1\right) \qquad (6.63)$$

Close to the sun (6.63) means that

$$v_\varphi \approx r\Omega_\odot$$

i.e., rigid rotation with the sun; on the other hand, at large distances from the sun

$$v_\varphi \approx \frac{r_A^2\Omega_\odot}{r}$$

which expresses the conservation of angular momentum from the Alfvénic point outwards. The precise value of r_A is not known. It was believed to be typically $20R_\odot$, extrapolation of Helios results suggest a mean value of $12R_\odot$.

Bernoulli's equation

In order to obtain Bernoulli's equation we have to work on the radial component of the equation of motion (6.56). What we have to do now is to combine and manipulate the different terms of this equation so that they are written as derivatives. This is rather tricky and do not waste your time and energy on trying to memorize the various steps. The writer does not know them by heart and is not all worried by that fact. The outcome is Bernoulli's equation:

$$\frac{v_r^2}{2} + \frac{1}{2}(v_\varphi - r\Omega_\odot)^2 - \frac{GM_\odot}{r} + \frac{\gamma}{\gamma - 1}\frac{p}{\rho} - \frac{\Omega_\odot^2 r^2}{2} = E \qquad (6.64)$$

The assumption of polytropic flow has been used as can be seen from the term $\gamma p/((\gamma - 1)\rho)$. As before we define a Bernoulli's function. Here it is defined as

$$H(r, v_r, v_\varphi, \rho, p) = \frac{v_r^2}{2} + \frac{1}{2}(v_\varphi - r\Omega)^2 - \frac{GM_\odot}{r} + \frac{\gamma}{\gamma - 1}\frac{p}{\rho} - \frac{\Omega^2 r^2}{2} \qquad (6.65)$$

v_r, v_φ, ρ, p are 4 dependent variables. It is possible to express 3 out of these 4 dependent variables in terms of the remaining fourth quantity. For example we can express v_r, v_φ, p in terms of ρ so that $H(r, v_r, v_\varphi, \rho, p)$ is reduced to $H(r, \rho)$. Bernoulii's equation is then

$$H(r, \rho) = E$$

The solutions are the level curves of the function $H(r, \rho)$. The corresponding differential equation is

$$\frac{\partial H}{\partial \rho}\frac{d\rho}{dr} + \frac{\partial H}{\partial r} = 0 \tag{6.66}$$

The critical points of $H(r, \rho)$ are determined by

$$\frac{\partial H}{\partial \rho} = 0, \quad \frac{\partial H}{\partial r} = 0.$$

Moreover

$$\frac{\partial H}{\partial \rho} = 0$$

determines the singularities of (6.66). This condition combined with the requirement that $d\rho/dr$ is finite, means that the singularities of (6.66) correspond to the critical points of $H(r, \rho)$. The next step is to rewrite the various terms of $H(r, v_r, v_\varphi, \rho, p)$ so that they only contain r and ρ. We need to go through a bit of straightforward but tedious algebra. So, the reader is again warned that he/she is forbidden to memorize these results. Useful intermediate results are

$$\begin{aligned}
T_1 &= \frac{v_r^2}{2} = \frac{C_1^2}{2\rho^2 r^4}, \\
T_2 &= \frac{1}{2}(v_\varphi - r\Omega)^2 = \frac{1}{2}\left(\frac{v_r B_\varphi}{B_r}\right)^2 = \frac{1}{2}\Omega_\odot^2 r_A^2 \frac{(r_A/r - r/r_A)^2}{(1-\rho/\rho_A)^2}, \\
T_4 &= \frac{\gamma}{\gamma-1}\frac{p}{\rho} = \frac{\gamma}{\gamma-1}K\rho^{\gamma-1}
\end{aligned}$$

$\frac{\partial H}{\partial \rho} = 0$ defines the singularities of the differential equation and also (two of the three) critical points. Hence, we have to use the results for T_1, T_2, T_4 to compute $\frac{\partial H}{\partial \rho}$ or better to compute $\rho\frac{\partial H}{\partial \rho}$ as :

$$\begin{aligned}
\rho\frac{\partial H}{\partial \rho} &= \frac{-C_1^2}{\rho^2 r^4} - \Omega_\odot^2 r_A^2 \frac{(r_A/r - r/r_A)^2}{(1-\rho/\rho_A)^2}\frac{\rho/\rho_A}{(\rho/\rho_A - 1)} + \gamma K\rho^{\gamma-1} \\
&= -v_r^2 + \frac{v_r^2 v_{A,\varphi}^2}{v_r^2 - v_{A,r}^2} + v_S^2
\end{aligned}$$

The singularities of the differential equation are determined by

$$-v_r^2 + \frac{v_r^2 v_{A,\varphi}^2}{v_r^2 - v_{A,r}^2} + v_S^2 = 0$$

which we can rewrite as a quadratic equation for v_r^2

$$v_r^4 - v_r^2(v_A^2 + v_S^2) + v_S^2 v_{A,r}^2 = 0 \qquad (6.67)$$

The two solutions are

$$(v_r^2)_{1,2} = \frac{1}{2}\left\{(v_A^2 + v_S^2) \pm \sqrt{(v_A^2 + v_S^2)^2 - 4v_{A,r}^2 v_S^2}\right\} \qquad (6.68)$$

In order to understand the results of (6.68) we have to go back to the previous Chapter. In particular we have to go back to the discussion of magneto-acoustic waves. The frequencies of these waves are given by (5.28). (5.28) can be rewritten as

$$\sigma_{sl,f}^2 = \frac{k^2}{2}\left\{(v_A^2 + v_S^2) \pm \sqrt{(v_A^2 + v_S^2)^2 - 4k_x^2 v_A^2 v_S^2/k^2}\right\}$$

and the phase velocities of the slow and fast magneto-acoustic waves are

$$v_{ph,sl,f}^2 = \frac{1}{2}\left\{(v_A^2 + v_S^2) \pm \sqrt{(v_A^2 + v_S^2)^2 - 4k_x^2 v_A^2 v_S^2/k^2}\right\}$$

If we agree that $(k_x^2/k^2)v_A^2$ in the Cartesian system can be replaced with $v_{A,r}^2$ in the spherical system, then we can identify the velocities that correspond to the singularities of (6.66) as the phase speeds of the slow and fast magneto-acoustic waves. For that reason we denote these velocities as $v_{f,r}$ and $v_{sl,r}$. The two singularities of (6.66) occur at the two positions where v_r equals the speed of the slow wave and the fast wave respectively. Let us recall that we already found the Alfvén critical point from the φ-component of the equation of motion. Hence, the three basic MHD waves are instrumental in determining the topology of the magnetized wind. The magnetized wind has three critical points determined by the phase velocities of the three basic MHD waves. The critical points occur at the positions r_{sl}, r_A, r_f where the radial outflow velocity v_r equals the slow phase speed $v_{sl,r}$, the Alfvén phase speed $v_{A,r}$ and the fast phase speed $v_{f,r}$ respectively. In a non-magnetic plasma signals travel at the sonic speed. A singularity in the flow occurs when the velocity equals this characteristic speed. In a magnetized plasmas signals can travel at three different speeds corresponding to the three basic MHD waves. Singularities occur whenever the outflow velocity equals one of these characteristic speeds.

Differential equation for v_r

(6.66) is a differential equation for ρ. Using (6.68) we can rewrite the expression for $\rho\partial H/\partial\rho$ as

$$\rho\frac{\partial H}{\partial \rho} = -\frac{(v_r^2 - v_{sl,r}^2)(v_r^2 - v_{f,r}^2)}{(v_r^2 - v_{A,r}^2)}$$

The differential equation for v_r can be obtained by combining the continuity equation (6.23) and the differential equation for ρ (6.66) into

$$\rho\frac{\partial H}{\partial \rho}\frac{1}{v_r}\frac{dv_r}{dr} = \frac{\partial H}{\partial r} - \frac{2\rho}{r}\frac{\partial H}{\partial \rho}$$

or

$$D\frac{dv_r}{dr} = \frac{v_r}{r}N \qquad (6.69)$$

where

$$D = v_r^4 - v_r^2(v_A^2 + v_S^2) + v_S^2 v_{A,r}^2$$

$$N = (v_r^2 - v_{A,r}^2)(2v_S^2 + v_\varphi^2 - G\mathcal{M}_\odot/r) + 2v_r v_\varphi v_{A,r} v_{A,\varphi}$$

Results for the Weber-Davis model

Obtaining a full solution requires numeric calculation. The conditions

$$D(r_{sl}, v_{sl,r}) = 0, \quad D(r_f, v_{f,r}) = 0, \quad N(r_{sl}, v_{sl,r}) = 0$$

$$N(r_f, v_{f,r}) = 0, \quad H(r_{sl}, v_{sl,r}) = E, \quad H(r_f, v_{f,r}) = E$$

are six coupled transcendental solutions, which must be solved simultaneously for the six quantities $v_r(r_0), v_\varphi(r_0), r_{sl}, v_{sl,r}, r_f, v_{f,r}$. It is a well-posed problem, but it is no piece-of-cake. The attentive reader is wondering whether the solution should not also pass through the Alfvén point, which is a critical point of the $\varphi-$ component. Of course, it should. It turns out that any solution going through the slow and fast critical points automatically goes through the Alfvén point. The topology of the WD solutions is illustrated on Fig. 6.7. This figure is borrowed from C1994 "Large-scale Dynamics of the Solar Wind", unpublished class notes, 1994. The solutions are given for the present-day Sun and for a strongly magnetized rapidly rotating Sun. The latter rotates 25 times faster and has a 25 times stronger surface magnetic field strength than the present sun. In C1994's notation the sound velocity is c_s. Its variation is shown as the dotted line. Its value at r_0 is denoted as c_{s0}. The triangle, solid dot and the diamond indicate the positions of the slow, Alfvén and fast critical point respectively. For the present-day Sun the Alfvén and fast critical point almost coincide. They cannot be distinguished on Fig. 6.7. Fig. 6.8 shows the variation of $v_{f,r}, v_{A,r}, v_r, v_S, v_{sl,r}$ with distance r. The radial distance r is expressed in the unit $r_0 = 1.25R_\odot$. The velocities are measured with the sound velocity at r_0 as unit.

- The topology of the Weber-Davis wind is a lot more complex than that of the Parker wind, or that of the rotating wind.
- There is one and only one accelerating solution that is subsonic at r_0 and connects to $r \to \infty$. It is the solution that passes through the three critical points.
- There is no longer any allowed fully subsonic solution.
- The slow magnetosonic critical point is the true sonic point for the wind solution. At the position r_{sl}, the fluid velocity is only slightly less than the sound velocity. This is Parker's critical point displaced slightly because the sound velocity is only slightly larger than the phase velocity of the slow magnetosonic wave as can be seen Fig. 6.8
- For the solar parameters the Alfvén critical point r_A and the fast critical point r_f nearly coincide as can be seen both on Fig. 6.7 and on Fig. 6.8.
- The Weber-Davis model does a good job at reproducing the observed properties of the low speed streams at 1 AU, in the sense that the predictions for the flow speed and the particle density are within the observed fluctuations. This is largely due to the polytropic approximation.

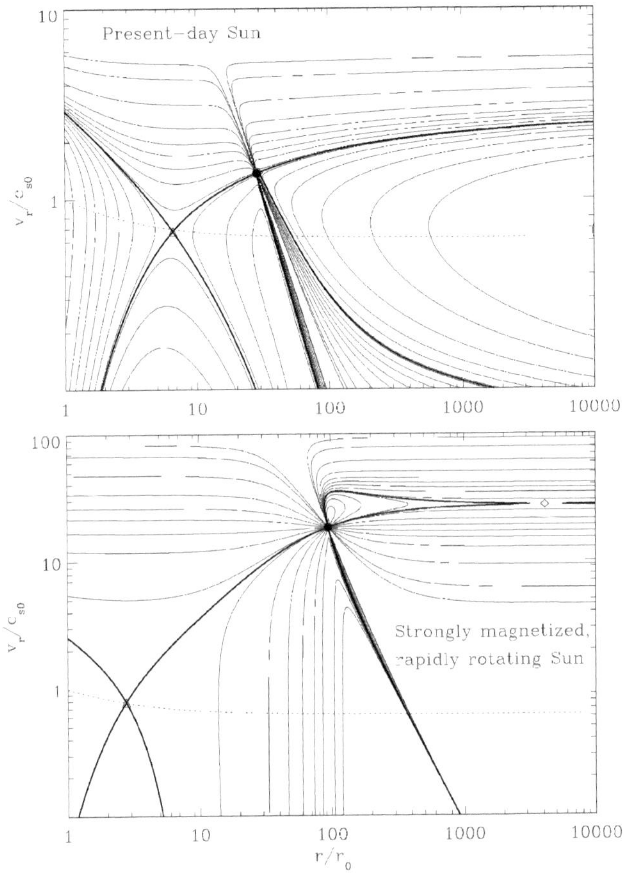

Figure 6.7: *Topology of the Weber-Davis solution for the Sun. (C1994.)*

- The Weber-Davis model predicts flow speeds and particle densities almost identical to the non-rotating unmagnetized polytropic models. The magnetic and centrifugal contributions to the acceleration are, everywhere in the flow, much smaller than the thermal acceleration.
- In the Weber-Davis model there is no significant acceleration beyond 1 AU and $\lim_{r\to\infty} v_r =$ constant. This is mostly a consequence of the polytropic approximation.
- The Weber-Davis model predicts values for $v_r, v_\varphi, B_r, B_\varphi$. At all distances $v_\varphi << v_r$. Values of v_r, v_φ can be translated into a deflection angle between the flow direction and

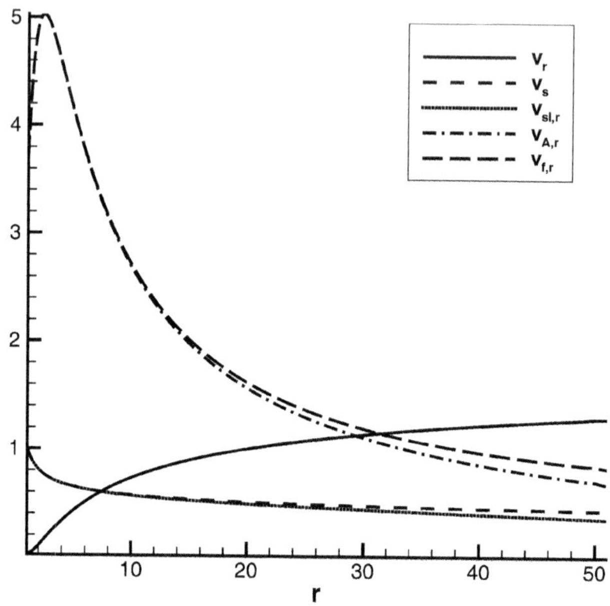

Figure 6.8: *The solution for the Weber-Davis wind model for v_r.*

the radial direction. The Weber-Davis model also yields directly a value for the angle Ψ between the direction of the magnetic field lines and the radial direction.

The effect of the mass loss due to the wind on the internal structure of the sun is negligible. Over the course of its evolution on the main sequence the Sun will loose a total mass fraction

$$\frac{\Delta\mathcal{M}_\odot}{\mathcal{M}_\odot} = \frac{d\mathcal{M}_\odot}{dt} \times \tau_{MS}$$

where $\tau_{MS} = 10^{10}$yr is the main-sequence life time of the sun. Once $v_r(r_0)$ is known, the mass loss rate can be computed:

$$\frac{d\mathcal{M}_\odot}{dt} = 2.4 \times 10^{-14}\mathcal{M}_\odot/yr$$

Hence

$$\frac{\Delta\mathcal{M}_\odot}{\mathcal{M}_\odot} \approx 2 \times 10^{-4}$$

This is too small to cause significant structural changes in the solar interior.

As for the total angular momentum carried away by the wind, it turns out that the specific angular momentum carried away by the gas is more than an order of magnitude smaller than the torque density associated with the magnetic field. The sum of the two is constant. Under the assumption that the solution is valid over the whole sphere, the loss of angular momentum is

$$\frac{dJ_\odot}{dt} = \frac{d\mathcal{M}_\odot}{dt} r_A^2 \Omega_\odot = 10^{26} Nm.$$

Assuming, for the sake of argument, that the Sun is rotating as a solid body and that its moment of inertia does not differ too greatly from that of a constant density sphere of the same mass and radius, we can write

$$J_\odot = \Omega_\odot I_\odot = \Omega_\odot \left(\frac{2}{5}\mathcal{M}_\odot R_\odot^2\right) \approx 10^{49} Nm/s$$

The time scale for angular momentum loss is $\tau_J = (1/J_\odot)(dJ_\odot/dt)$. With the numerical values for $J_\odot$ and $dJ_\odot/dt$ we get

$$\tau_J \approx 10^{23} s \approx 10^{10} yr$$

This is comparable to the main-sequence lifetime of the Sun, indicating that angular momentum loss by a magnetic wind can alter the rotation rate of solar-type stars as they evolve on the main sequence. However, the situation is actually more complicated than this timescale estimate may suggest. Simulations show that the solutions depend sensitively on τ and $\beta(r_0)$. In addition, the assumption that the solution in the equatorial plane is a realistic approximation of the wind over the whole sphere is not confirmed by recent 2-D computations. Anyhow, if the Sun rotated more rapidly and/or had a stronger magnetic field in the past, the angular momentum loss was probably larger than estimated here.

6.8 Recapitulation

"Has anything escaped me?" I asked, with some self-importance.
"I trust that there is nothing of consequence which I have overlooked?"
"I am afraid, my dear Watson, that most of your conclusions were erroneous."
Conversation of Sherlock Holmes with Dr. Watson
The Hound of the Baskervilles
A. Conan Doyle

- In this Chapter we have studied the solar wind, which has evolved from a concept originally received with a lot of scepticism into a fundamental phenomenon of solar physics and astrophysics.

- We have taken the historic route. Our starting point was a static conductive model for the solar corona. We found that it was impossible to embed this hot corona in the interstellar medium under pressure equilibrium. From then on we became dedicated followers of Bernoulli. His equation tells us that there is a transfer of enthalpy (random motions) to kinetic energy (directed motions) so that the hot corona cannot be static but must expand continuously into the near vacuum between the stars.

- We then became rocket scientists and looked at how a flow can be accelerated from subsonic to supersonic velocities in a de Laval nozzle. This acceleration is realized by the transonic solution. The principle on which the acceleration is based in the de Laval nozzle is the same as that in the solar wind: the conversion of enthalpy into kinetic energy. In the rocket engine the walls provide the confinement; the critical point is located in the throat of the nozzle leading to supersonic expansion.

- The static model of the solar corona is now replaced by a model with a radial outflow. The analysis of the de Laval nozzle can be repeated with little modifications. Gravity confines the plasma, the critical point is the sonic point and the transonic solution provides a wind that is subsonic at the solar surface and is supersonic beyond the critical point. Surprisingly enough the remarkably simple isothermal Parker model predicts the high supersonic velocities observed at the Earth's orbit.

- The next step is to include rotation in the analysis. Life becomes more complicated because we loose spherical symmetry. We downsize our ambition and content ourselves with solutions in the equatorial plane. The effect due to rotation is always small for the sun. In particular a non-magnetized wind does not transport a substantial amount of angular momentum.

- The final step is to include a magnetic field in the analysis. In spite of the fact that the energy contained in the magnetic field is much smaller than the rotational energy and the gravitational energy of the sun, the magnetic field is an important player. Here also we content ourselves with solutions in the equatorial plane.

- The Archimedean spiral of the magnetic field lines in the solar wind is found to be a direct consequence of flux freezing. A crucial result is that the angular momentum carried away by a magnetized wind is equal to the angular momentum carried away a non-magnetized wind flowing strictly radially and co-rotating to the Alfvén radius. The effective lever arm used by the wind to brake the Sun's rotation is the Alfvén radius resulting in a loss of angular momentum that is about 400 times larger than in a non-magnetized wind.

- The magnetic wind solution is now characterized by 3 critical points corresponding to the positions where the radial component of the outflow velocity is equal to the speeds of the 3 basic MHD waves. Weber and Davis were the first, way back in the sixties, to construct a 2-D polytropic model of magnetized rotating wind.

6.9 Problems

"Elementary, my dear Watson," the great detective said quietly, ···
"Entirely elementary."
Sherlock Holmes to Dr. Watson
The Hound of the Baskervilles
A. Conan Doyle

1. A comet's ion tail deviates by 4° from the radial direction. Its transverse orbital velocity is 30 km/s. What is the speed of the solar wind? Convince yourself that this speed is supersonic!

2. Derive Bernoulli's equation for a static isothermal equilibrium and use this equation to find density as function of r.

3. Replace the assumption that the coronal plasma is isothermal with the assumption that there is a a polytropic relation between pressure and density and determine the static polytropic corona model.

4. Investigate the variation of $\rho(r)$ as function of r in Chapman's coronal model and show that density first decreases, attains an absolute minimum and subsequently monotonically increases towards $+\infty$.

5. Go back to the relation between pressure and temperature for a static coronal model (6.9) and find out how fast $T(r)$ should go to zero for $r \to \infty$ in order that $p_\infty = 0$.

6. Use
$$\frac{\gamma}{(\gamma-1)}\frac{\overline{\mathcal{R}}}{\overline{\mu}}T_0 - \frac{G\mathcal{M}_\odot}{r_0} \approx \frac{v_\infty^2}{2}$$
with $T_0 = 1 \times 10^6$K and $r_0 = 1.2R_\odot$ to compute v_∞.

7. Recall that we have used the continuity equation
$$\sigma(x)v_x(x) = \text{constant}$$
for the study of the behaviour of incompressible flow in a nozzle. However, for compressible flow in a nozzle we have used the following non-linear first order ordinary differential equation for v_x:
$$\frac{1}{v_x}\left(\frac{v_x^2}{v_{S,i}^2} - 1\right)\frac{dv_x}{dx} = \frac{1}{\sigma}\frac{d\sigma}{dx}.$$
Find the corresponding differential equation for v_x for an incompressible flow (this is really easy, all you have to do is to differentiate the continuity equation!). Compare the differential equations for v_x for incompressible and compressible flow. Explain (or rather note) that you can obtain the differential equation for incompressible flow as

a special case of that for compressible flow by taking the limit $v_S^2 \to \infty$. This should give you a good intuitive understanding of the assumption of incompressibility. In an incompressible fluid sonic information is transmitted at infinite speed and all flows are necessarily subsonic.

8. Assume that the flow in the de Laval nozzle is isothermal. This is not a very good approximation but the intention is to have a Cartesian model that is as close as possible to Parker's isothermal wind model. Determine the following Bernoulli's function

$$H(x, v_x) = \frac{v_x^2}{v_S^2} - \ln \frac{\sigma^2}{\sigma(x_0)^2} - \ln \frac{v_x^2}{v_S^2}.$$

Determine its critical point and find that it is the sonic point $x = x_s, \; v_x = v_S$. Show that the critical point is a saddle point of the function $H(x, v_x)$.

9. Look again at an isothermal flow in a nozzle with cross-section

$$\sigma(x) = x^2 \exp \frac{2L}{x}$$

in the interval $0.01L \le x \le 10L$. Show that this nozzle can be used for the acceleration of subsonic flow at $x = 0.01L$ to supersonic flow at $x = 10L$. Where does the transition from subsonic to supersonic flow take place? Take $v_S = 1$ and $L = 1$ and find the transonic solution with $dv_x/dx > 0$, in particular look at the value of v_x at $x = 0.01$ and $x = 10$!

10. Compute r_s for a coronal temperature $T_0 = 10^6 K$. The sound speed is $v_{S,i} = (\overline{\mathcal{R}}T/\overline{\mu})^{1/2} \approx 10^5 m\ s^{-1}$. Compare your value for r_s with the radius of the Earth's orbit $R_E \approx 215 R_\odot$.

11. The existence of a solar wind depends on the basal temperature of the corona. If the temperature is too low, the situation is essentially static. If the temperature is too high there is effectively no throat. You can show that r_s is a monotonically decreasing function of T. If $r_s < r_0 \; (= 1.2R_\odot)$ then there is no sonic point and also no wind. Determine the critical temperature T_{cr} above which supersonic expansion cannot occur.

12. Determine the cross-section $\sigma(r)$ of the rocket engine that corresponds to the solar wind or in other words determine $\sigma(r)$ so that

$$\frac{1}{\sigma}\frac{d\sigma}{dr} = \frac{2}{r}\left(1 - \frac{G\mathcal{M}_\odot/2v_S^2}{r}\right).$$

13. Look at the second version of Bernoulli's function for an isothermal radial outflow

$$H(r, v_r) = \frac{v_r^2}{v_{S,i}^2} - \ln \frac{v_r^2}{v_{S,i}^2} - \ln \frac{r^4}{r_s^4} - 4\frac{r_s}{r}.$$

Determine its critical point and find that it is our sonic point $x = x_s \;\wedge\; v_x = v_{S,i}$. Show that the critical point is a saddle point of the function $H(r, v_r)$.

14. Look at the accelerating transonic solution for isothermal radial outflow. Determine the asymptotic behaviour of v_r and ρ for $r \to +\infty$. How can you explain your result for v_r?

15. Look at the subsonic solutions for isothermal radial outflow (i.e. the so called solar breeze!) Suppose you do not know about the satellite data on the supersonic velocities of the solar wind at the earth's orbit. Find another reason for discarding the subsonic solutions! Look at the asymptotic behaviour of v_r and ρ for $r \to +\infty$.

> "I don't mean to deny that the evidence is in some ways very strong in favour of your theory I only wish to point out that there are other theories possible."
> Sherlock Holmes
> *Adventure of the Norwood Builder*
> A. Conan Doyle

Eugene N. Parker (1927-)
Eugene N. After receiving a B.S. degree from Michigan State College in 1948 and a Ph.D. from the California Institute of Technology in 1951, he held various positions at the University of Utah from 1951 to 1955. In 1955, he joined the University of Chicago, where John Simpson and others were beginning to challenge the then-prevalent concept of interplanetary space being largely empty, traversed by a few fast-moving particles. With only a few in situ observations in the immediate neighborhood of the Earth, theorists had to rely on a variety of ambiguous observable geophysical and astronomical phenomena. In 1958, Parker published his theory of the solar wind, in which the solar corona expands supersonically to the outer reaches of the solar system, now the foundation of modern solar-terrestrial research and solar-planetary relationships. Prior to Parker's work, the existence of a continuous, but slight, flow of particles from the Sun was suggested by a variety of circumstantial evidence. The idea of such a vigorous, dense, and dynamically complex outflow was so radical for its time that it drew criticism and disbelief from most of the scientific community. In 1960, Soviet scientists reported suggestive observations by Lunik 2; in 1961, Explorer 10 data provided convincing confirmation, followed by Mariner 2 observations in 1962 between the orbits of Earth and Venus. Studying the detailed structure and dynamics of the solar wind animated much of the scientific exploration of the inner solar system during the 1960s, and the concept, extended to other stars, became one of the most important foundations of modern astrophysics.

Joseph N. Tatarewicz
University of Maryland Baltimore County Catonsville, Maryland

"And now, my dear Watson, we have had some weeks of severe work, and for one evening, I think, we may turn our thoughts into more pleasant channels."
Sherlock Holmes to Dr. Watson
The Hound of the Baskervilles
A. Conan Doyle

Physical constants

Electron rest mass	m_e	9.1094×10^{-31} kg
Proton rest mass	m_p	1.6726×10^{-27} kg
Proton-electron mass ratio	m_p/m_e	1836
Electron charge	e	1.6022×10^{-19} C
Permeability of free space	μ_0	$4\pi \times 10^{-7}$H m^{-1}
Permittivity of free space	ϵ_0	8.854×10^{-12}F m^{-1}
Velocity of light	c	2.9979×10^8m s^{-1}
Gravitational constant	G	6.671×10^{-11}m^{3}s^{-2}kg^{-1}
Boltzmann's constant	k_B	$1.3807 \times 10^{-23} J$K^{-1}
	k_B/m_p	8.2548×10^3m^{2}s^{-2}K^{-1}
1 eV Temperature		1.1605×10^4 K

Solar properties

Mass	$\mathcal{M}_\odot$	1.99×10^{30} kg
Radius	$R_\odot$	6.96×10^8 m
Escape velocity	$v_{esc,\odot}$	618 km s^{-1}
Surface gravity	$g_\odot$	274 m s^{-2}
Equatorial rotation period	$P_{rot,\odot}$	26 days
Distance to earth	1 AU	1.50×10^{11}m $= 215 R_\odot$

Frequencies and speeds

Electron thermal speed	$v_{t,e}$	$(k_B T_e/m_e)^{1/2}$
Ion thermal speed	$v_{t,i}$	$(k_B T_p/m_i)^{1/2}$
Electron plasma frequency	$\omega_{p,e}$	$(n_e e^2/m_e \epsilon_0)^{1/2}$
Ion plasma frequency	$\omega_{p,i}$	$(n_i Z^2 e^2/m_i \epsilon_0)^{1/2}$
Debey length	λ_D	$(\epsilon_0 k_B T_e/n_e e^2)^{1/2}$
		$= v_{t,e}/\omega_{p,e}$
Electron gyro-frequency	$\omega_{c,e}$	eB/m_e
Ion gyro-frequency	$\omega_{c,i}$	ZeB/m_i
Electron gyro-radius at thermal speed	$r_{L,t,e}$	$v_{t,e}/\omega_{c,e}$
Ion gyro-radius at thermal speed	$r_{L,t,i}$	$v_{t,i}/\omega_{c,i}$
Sound speed	v_S	$(\gamma p/\rho)^{1/2}$
Alfvén speed speed	v_A	$(B^2/\mu\rho)^{1/2}$

GPSR Compliance

The European Union's (EU) General Product Safety Regulation (GPSR) is a set of rules that requires consumer products to be safe and our obligations to ensure this.

If you have any concerns about our products, you can contact us on ProductSafety@springernature.com

In case Publisher is established outside the EU, the EU authorized representative is:

Springer Nature Customer Service Center GmbH
Europaplatz 3
69115 Heidelberg, Germany

Batch number: 10583998

Printed by Printforce, the Netherlands